MEMORIAL VOLUME
FOR
Y. NAMBU

MEMORIAL VOLUME FOR Y. NAMBU

Editors

Lars Brink
Chalmers University of Technology, Sweden

Lay Nam Chang
Virginia Tech, USA

Moo-Young Han
Duke University, USA

Kok Khoo Phua
Nanyang Technological University, Singapore

NEW JERSEY • LONDON • SINGAPORE • BEIJING • SHANGHAI • HONG KONG • TAIPEI • CHENNAI • TOKYO

Published by

World Scientific Publishing Co. Pte. Ltd.
5 Toh Tuck Link, Singapore 596224
USA office: 27 Warren Street, Suite 401-402, Hackensack, NJ 07601
UK office: 57 Shelton Street, Covent Garden, London WC2H 9HE

Library of Congress Cataloging-in-Publication Data
Names: Brink, Lars, 1945– editor. | Chang, L. N. (Lay Nam), editor. | Han, M. Y., editor. | Phua, K. K., editor. | Nambu, Yoichiro, 1921–2015, honouree.
Title: Memorial volume for Y. Nambu / editors: Lars Brink (Chalmers University of Technology, Sweden), Lay Nam Chang (Virginia Tech, USA), Moo-Young Han (Duke University, USA), Kok Khoo Phua (NTU, Singapore).
Other titles: Memorial volume for Yoichiro Nambu
Description: Singapore ; Hackensack, NJ : World Scientific Publishing Co. Pte. Ltd., [2016] | 2016 | Includes bibliographical references.
Identifiers: LCCN 2016013313| ISBN 9789813108318 (hardcover ; alk. paper) | ISBN 9813108312 (hardcover ; alk. paper) | ISBN 9789813108325 (pbk ; alk. paper) | ISBN 9813108320 (pbk ; alk. paper)
Subjects: LCSH: Physics. | Particles (Nuclear physics) | Field theory (Physics) | Nambu, Yoichiro, 1921-2015.
Classification: LCC QC6.2 .M46 2016 | DDC 530--dc23
LC record available at http://lccn.loc.gov/2016013313

British Library Cataloguing-in-Publication Data
A catalogue record for this book is available from the British Library.

Cover image credit: Betsy Devine

Desk Editor: Ng Kah Fee

Typeset by Stallion Press
Email: enquiries@stallionpress.com

Printed in Singapore

Contents

Foreword

Yoichiro Nambu (1921–2015) was one of the physics giants of the second half of the 20th century, which was the golden age of symmetry, gauge fields, and cosmology. The impact of his many pioneering works was enormous in many diverse areas of physics. This is particularly the case in his introduction of the concept of spontaneously broken symmetries, which has spawned breakthroughs in particle physics, condensed matter physics, hydrodynamics, among other areas.

In particle physics, for example, the concept, together with the recognition of a new symmetry for quarks and gluons, laid the cornerstone for the Standard Model of Particles and Forces. They led to the discovery of the Higgs particle in 2012, and to the formulation of Quantum Chromodynamics. It would not be an overstatement to say that without Nambu's incisive works, today's Standard Model would have taken a much longer time to develop.

With deep insights and penetrating vision, he saw things that others could not see and foretold the future developments. To quote Bruno Zumino, "He is always 10 years ahead of us, so I tried to understand his works in order to contribute to a new area which will flourish 10 years later. Contrary to my expectations, however, it took me 10 years to understand them."

As a person, Nambu was one of the most easily accessible and kind mentor for such a giant physicist whom we all fondly remember as a great friend who was unassuming, considerate, and modest.

If Nambu is to be characterized in one short phrase, it is that he was a person of humble modesty and quiet dignity.

The Editors and World Scientific Publishing take great pleasure in presenting this volume in which we have collected articles by friends, colleagues and former associates and students celebrating the life and science of Yoichiro Nambu. We hope and believe that the readers would find this volume not only very useful but also cherish it as a memorial to a man many refer to as "Nambu the seer."

Lars Brink
Lay Nam Chang
Moo-Young Han
Kok Khoo Phua
March 2016

Chapter 1

Effective field theory, past and future

Steven Weinberg

Theory Group, Department of Physics,
University of Texas, Austin, TX, 78712
weinberg@physics.utexas.edu

This is a written version of the opening talk at the 6th International Workshop on Chiral Dynamics, at the University of Bern, Switzerland, July 6, 2009, to be published in the proceedings of the Workshop. In it, I reminisce about the early development of effective field theories of the strong interactions, comment briefly on some other applications of effective field theories, and then take up the idea that the Standard Model and General Relativity are the leading terms in an effective field theory. Finally, I cite recent calculations that suggest that the effective field theory of gravitation and matter is asymptotically safe.

I have been asked by the organizers of this meeting to "celebrate 30 years" of a paper[1] on effective field theories that I wrote in 1979. I am quoting this request at the outset, because in the first half of this talk I will be reminiscing about my own work on effective field theories, leading up to this 1979 paper. I think it is important to understand how confusing these things seemed back then, and no one knows better than I do how confused I was. But I am sure that many in this audience know more than I do about the applications of effective field theory to the strong interactions since 1979, so I will mention only some early applications to strong interactions and a few

[1]S. Weinberg, *Physica A* **96**, 327 (1979).

applications to other areas of physics. I will then describe how we have come to think that our most fundamental theories, the Standard Model and General Relativity, are the leading terms in an effective field theory. Finally, I will report on recent work of others that lends support to a suggestion that this effective field theory may actually be a fundamental theory, valid at all energies.

It all started with current algebra. As everyone knows, in 1960 Yoichiro Nambu had the idea that the axial vector current of beta decay could be considered to be conserved in the same limit that the pion, the lightest hadron, could be considered massless.[2] This assumption would follow if the axial vector current was associated with a spontaneously broken approximate symmetry, with the pion playing the role of a Goldstone boson.[3] Nambu used this idea to explain the success of the Goldberger–Treiman formula[4] for the pion decay amplitude, and with his collaborators he was able to derive formulas for the rate of emission of a single low energy pion in various collisions.[5] In this work it was not necessary to assume anything about the nature of the broken symmetry — only that there was some approximate symmetry responsible for the approximate conservation of the axial vector current and the approximate masslessness of the pion. But to deal with processes involving more than one pion, it was necessary to use not only the approximate conservation of the current but also the current commutation relations, which of course do depend on the underlying broken symmetry. The technology of using these properties of the currents, in which one does not use any specific Lagrangian for the strong interactions, became known as current algebra.[6] It scored a dramatic success in the derivation

[2]Y. Nambu, *Phys. Rev. Lett.* **4**, 380 (1960).
[3]J. Goldstone, *Nuovo Cimento* **9**, 154 (1961); Y. Nambu and G. Jona-Lasinio, *Phys. Rev.* **122**, 345 (1961); J. Goldstone, A. Salam, and S. Weinberg, *Phys. Rev.* **127**, 965 (1962).
[4]M. L. Goldberger and S. B. Treiman, *Phys. Rev.* **111**, 354 (1956).
[5]Y. Nambu and D. Lurie, *Phys Rev.* **125**, 1429 (1962); Y. Nambu and E. Shrauner, *Phys. Rev.* **128**, 862 (1962).
[6]The name may be due to Murray Gell-Mann. The current commutation relations were given in M. Gell-Mann, *Physics* **1**, 63 (1964).

of the Adler–Weisberger sum rule[7] for the axial vector beta decay coupling constant g_A, which showed that the current commutation relations are those of $SU(2) \times SU(2)$.

When I started in the mid-1960s to work on current algebra, I had the feeling that, despite the success of the Goldberger–Treiman relation and the Adler–Weisberger sum rule, there was then rather too much emphasis on the role that the axial vector current plays in weak interactions. After all, even if there were no weak interactions, the fact that the strong interactions have an approximate but spontaneously broken $SU(2) \times SU(2)$ symmetry would be a pretty important piece of information about the strong interactions.[8] To demonstrate the point, I was able to use current algebra to derive successful formulas for the pion–pion and pion–nucleon scattering lengths.[9] When combined with a well-known dispersion relation[10] and the Goldberger–Treiman relation, these formulas for the pion–nucleon scattering lengths turned out to imply the Adler–Weisberger sum rule.

In 1966 I turned to the problem of calculating the rate of processes in which arbitrary numbers of low energy massless pions are emitted in the collision of other hadrons. This was not a problem that urgently needed to be solved. I was interested in it because a year earlier I had worked out simple formulas for the rate of emission of arbitrary numbers of soft gravitons or photons in any collision,[11] and I was curious whether anything equally simple could be said about soft pions. Calculating the amplitude for emission of several soft pions by use of the technique of current

---

[7]S. L. Adler, *Phys. Rev. Lett.* **14**, 1051 (1965); *Phys. Rev.* **140**, B736 (1965); W. I. Weisberger, *Phys. Rev. Lett.* **14**, 1047 (1965).

[8]I emphasized this point in my rapporteur's talk on current algebra at the 1968 "Rochester" conference; see *Proceedings of the 14th International Conference on High-Energy Physics*, p. 253.

[9]S. Weinberg, *Phys. Rev. Lett.* **17**, 616 (1966). The pion–nucleon scattering lengths were calculated independently by Y. Tomozawa, *Nuovo Cimento* **46A**, 707 (1966).

[10]M. L. Goldberger, Y. Miyazawa, and R. Oehme, *Phys. Rev.* **99**, 986 (1955).

[11]S. Weinberg, *Phys. Rev.* **140**, B516 (1965).

algebra turned out to be fearsomely complicated; the non-vanishing commutators of the currents associated with the soft pions prevented the derivation of anything as simple as the results for soft photons or gravitons, except in the special case in which all pions have the same charge.[12]

Then some time late in 1966 I was sitting at the counter of a café in Harvard Square, scribbling on a napkin the amplitudes I had found for emitting two or three soft pions in nucleon collisions, and it suddenly occurred to me that these results looked very much like what would be given by lowest order Feynman diagrams in a quantum field theory in which pion lines are emitted from the external nucleon lines, with a Lagrangian in which the nucleon interacts with one, two, and more pion fields. Why should this be? Remember, this was a time when theorists had pretty well given up the idea of applying specific quantum field theories to the strong interactions, because there was no reason to trust the lowest order of perturbation theory, and no way to sum the perturbation series. What was popular was to exploit tools such as current algebra and dispersion relations that did not rely on assumptions about particular Lagrangians.

The best explanation that I could give then for the field-theoretic appearance of the current algebra results was that these results for the emission of n soft pions in nucleon collisions are of the minimum order, G_π^n, in the pion–nucleon coupling constant G_π, except that one had to use the exact values for the collision amplitudes without soft pion emission, and divide by factors of the axial vector coupling constant $g_A \simeq 1.2$ in appropriate places. Therefore any Lagrangian that satisfied the axioms of current algebra would have to give the same answer as current algebra in lowest order perturbation theory, except that it would have to be a field theory in which soft pions were emitted only from external lines of the diagram for the nucleon collisions, for only then one would know how to put in the correct factors of g_A and the correct nucleon collision amplitude.

[12] S. Weinberg, *Phys. Rev. Lett.* **16**, 879 (1966).

The time-honored renormalizable theory of nucleons and pions with conserved currents that satisfied the assumptions of current algebra was the "linear σ-model,"[13] with Lagrangian (in the limit of exact current conservation):

$$\begin{aligned}\mathcal{L} = &-\frac{1}{2}\left[\partial_\mu \vec{\pi}\cdot\partial^\mu\vec{\pi} + \partial_\mu\sigma\,\partial^\mu\sigma\right] \\ &-\frac{m^2}{2}(\sigma^2+\vec{\pi}^2) - \frac{\lambda}{4}(\sigma^2+\vec{\pi}^2)^2 \\ &-\bar{N}\gamma^\mu\partial_\mu N - G_\pi \bar{N}(\sigma + 2i\gamma_5\vec{\pi}\cdot\vec{t}\,)N\,, \end{aligned} \tag{1}$$

where N, $\vec{\pi}$, and σ are the fields of the nucleon doublet, pion triplet, and a scalar singlet, and $\vec{t}$ is the nucleon isospin matrix (with $\vec{t}^{\,2} = 3/4$). This Lagrangian has an $SU(2)\times SU(2)$ symmetry (equivalent as far as current commutation relations are concerned to an $SO(4)$ symmetry), that is spontaneously broken for $m^2 < 0$ by the expectation value of the σ field, given in lowest order by $\langle\sigma\rangle = F/2 \equiv \sqrt{-m^2/\lambda}$, which also gives the nucleon a lowest order mass $2G_\pi F$. But with a Lagrangian of this form soft pions could be emitted from internal as well as external lines of the graphs for the nucleon collision itself, and there would be no way to evaluate the pion emission amplitude without having to sum over the infinite number of graphs for the nucleon collision amplitude.

To get around this obstacle, I used the chiral $SO(4)$ symmetry to rotate the chiral four-vector into the fourth direction

$$(\vec{\pi},\sigma)\mapsto(0,\sigma'),\qquad \sigma' = \sqrt{\sigma^2+\vec{\pi}^2}, \tag{2}$$

with a corresponding chiral transformation $N\mapsto N'$ of the nucleon doublet. The chiral symmetry of the Lagrangian would result in the pion disappearing from the Lagrangian, except that the matrix of the

[13] J. Bernstein, S. Fubini, M. Gell-Mann, and W. Thirring, *Nuovo Cimento* **17**, 757 (1960); M. Gell-Mann and M. Lévy, *Nuovo Cimento* **16**, 705 (1960); K. C. Chou, *Soviet Physics JETP* **12**, 492 (1961). This theory, with the inclusion of a symmetry-breaking term proportional to the σ field, was intended to provide an illustration of a "partially conserved axial current," that is, one whose divergence is proportional to the pion field.

rotation (2) necessarily, like the fields, depends on spacetime position, while the theory is only invariant under spacetime-*independent* chiral rotations. The pion field thus reappears as a parameter in the $SO(4)$ rotation (2), which could conveniently be taken as

$$\vec{\pi}' \equiv F\vec{\pi}/[\sigma + \sigma'] \,. \tag{3}$$

But the rotation parameter $\vec{\pi}'$ would not appear in the transformed Lagrangian if it were independent of the spacetime coordinates, so wherever it appears it must be accompanied with at least one derivative. This derivative produces a factor of pion four-momentum in the pion emission amplitude, which would suppress the amplitude for emitting soft pions, if this factor were not compensated by the pole in the nucleon propagator of an external nucleon line to which the pion emission vertex is attached. Thus, with the Lagrangian in this form, pions of small momenta can only be emitted from external lines of a nucleon collision amplitude. This is what I needed.

Since σ' is chiral-invariant, it plays no role in maintaining the chiral invariance of the theory, and could therefore be replaced with its lowest-order expectation value $F/2$. The transformed Lagrangian (now dropping primes) is then

$$\begin{aligned}\mathcal{L} = &-\frac{1}{2}\left[1+\frac{\vec{\pi}^2}{F^2}\right]^{-2} \partial_\mu\vec{\pi}\cdot\partial^\mu\vec{\pi} \\ &-\bar{N}\left[\gamma^\mu\partial_\mu + G_\pi F/2 + i\gamma^\mu\left[1+\frac{\vec{\pi}^2}{F^2}\right]^{-1}\right. \\ &\times\left.\left[\frac{2}{F}\gamma_5\vec{t}\cdot\partial_\mu\vec{\pi} + \frac{2}{F^2}\vec{t}\cdot(\vec{\pi}\times\partial_\mu\vec{\pi})\right]\right]N \,. \end{aligned} \tag{4}$$

In order to reproduce the results of current algebra, it is only necessary to identify F as the pion decay amplitude $F_\pi \simeq 184$ MeV, replace the term $G_\pi F/2$ in the nucleon bilinear with the actual nucleon mass m_N (given by the Goldberger–Treiman relation as $G_\pi F_\pi/2g_A$), and replace the pseudovector pion–nucleon coupling $1/F$ with its actual value $G_\pi/2m_N = g_A/F_\pi$. This gives an effective

Lagrangian

$$\mathcal{L}_{\text{eff}} = -\frac{1}{2}\left[1+\frac{\vec{\pi}^2}{F_\pi^2}\right]^{-2} \partial_\mu\vec{\pi}\cdot\partial^\mu\vec{\pi}$$
$$- \bar{N}\left[\gamma^\mu\partial_\mu + m_N + i\gamma^\mu\left[1+\frac{\vec{\pi}^2}{F_\pi^2}\right]^{-1}\right.$$
$$\left.\times\left[\frac{G_\pi}{m_N}\gamma_5\vec{t}\cdot\partial_\mu\vec{\pi} + \frac{2}{F_\pi^2}\vec{t}\cdot(\vec{\pi}\times\partial_\mu\vec{\pi})\right]\right]N\,. \tag{5}$$

To take account of the finite pion mass, the linear sigma model also includes a chiral-symmetry breaking perturbation proportional to σ. Making the chiral rotation (2), replacing σ' with the constant $F/2$, and adjusting the coefficient of this term to give the physical pion mass m_π gives a chiral symmetry breaking term

$$\Delta\mathcal{L}_{\text{eff}} = -\frac{1}{2}\left[1+\frac{\vec{\pi}^2}{F_\pi^2}\right]^{-1} m_\pi^2\,\vec{\pi}^2\,. \tag{6}$$

Using $\mathcal{L}_{\text{eff}} + \Delta\mathcal{L}_{\text{eff}}$ in lowest order perturbation theory, I found the same results for low-energy pion–pion and pion–nucleon scattering that I had obtained earlier with much greater difficulty by the methods of current algebra.

A few months after this work, Julian Schwinger remarked to me that it should be possible to skip this complicated derivation, forget all about the linear σ-model, and instead infer the structure of the Lagrangian directly from the non-linear chiral transformation properties of the pion field appearing in (5).[14] It was a good idea. I spent the summer of 1967 working out these transformation properties, and what they imply for the structure of the Lagrangian.[15]

[14]For Schwinger's own development of this idea, see J. Schwinger, *Phys. Lett.* **24B**, 473 (1967). It is interesting that in deriving the effective field theory of goldstinos in supergravity theories, it is much more transparent to start with a theory with linearly realized supersymmetry and impose constraints analogous to setting $\sigma' = F/2$, than to work from the beginning with supersymmetry realized non-linearly, in analogy to Eq. (7); see Z. Komargodski and N. Seiberg, to be published.

[15]S. Weinberg, *Phys. Rev.* **166**, 1568 (1968).

It turns out that if we require that the pion field has the usual linear transformation under $SO(3)$ isospin rotations (because isospin symmetry is supposed to be not spontaneously broken), then there is a *unique* $SO(4)$ chiral transformation that takes the pion field into a function of itself — unique, that is, up to possible redefinition of the field. For an infinitesimal $SO(4)$ rotation by an angle ϵ in the $a4$ plane (where $a = 1, 2, 3$), the pion field π_b (labelled with a prime in Eq. (3)) changes by an amount

$$\delta_a \pi_b = -i\epsilon F_\pi \left[\frac{1}{2} \left(1 - \frac{\vec{\pi}^2}{F_\pi^2} \right) \delta_{ab} + \frac{\pi_a \pi_b}{F_\pi^2} \right]. \tag{7}$$

Any other field ψ, on which isospin rotations act with a matrix $\vec{t}$, is changed by an infinitesimal chiral rotation in the $a4$ plane by an amount

$$\delta_a \psi = \frac{\epsilon}{F_\pi} (\vec{t} \times \vec{\pi})_a \psi. \tag{8}$$

This is just an ordinary, though position-dependent, isospin rotation, so a non-derivative isospin-invariant term in the Lagrangian that does not involve pions, like the nucleon mass term $-m_N \bar{N} N$, is automatically chiral-invariant. The terms in Eq. (5):

$$-\bar{N} \left[\gamma^\mu \partial_\mu + \frac{2i}{F_\pi^2} \gamma^\mu \left[1 + \frac{\vec{\pi}^2}{F_\pi^2} \right]^{-1} \vec{t} \cdot (\vec{\pi} \times \partial_\mu \vec{\pi}) \right] N, \tag{9}$$

and

$$-i \frac{G_\pi}{m_N} \left[1 + \frac{\vec{\pi}^2}{F_\pi^2} \right]^{-1} \bar{N} \gamma^\mu \gamma_5 \vec{t} \cdot \partial_\mu \vec{\pi} N, \tag{10}$$

are simply proportional to the most general chiral-invariant nucleon–pion interactions with a single spacetime derivative. The coefficient of the term (9) is fixed by the condition that N should be canonically normalized, while the coefficient of (10) is chosen to agree with the conventional definition of the pion–nucleon coupling G_π, and is not directly constrained by chiral symmetry. The term

$$-\frac{1}{2} \left[1 + \frac{\vec{\pi}^2}{F^2} \right]^{-2} \partial_\mu \vec{\pi} \cdot \partial^\mu \vec{\pi} \tag{11}$$

is proportional to the most general chiral invariant quantity involving the pion field and no more than two spacetime derivatives, with a coefficient fixed by the condition that $\vec{\pi}$ should be canonically normalized. The chiral symmetry breaking term (6) is the most general function of the pion field without derivatives that transforms as the fourth component of a chiral four-vector. None of this relies on the methods of current algebra, though one can use the Lagrangian (5) to calculate the Noether current corresponding to chiral transformations, and recover the Goldberger–Treiman relation in the form $g_A = G_\pi F_\pi / 2m_N$.

This sort of direct analysis was subsequently extended by Callan, Coleman, Wess, and Zumino to the transformation and interactions of the Goldstone boson fields associated with the spontaneous breakdown of any Lie group G to any subgroup H.[16] Here, too, the transformation of the Goldstone boson fields is unique, up to a redefinition of the fields, and the transformation of other fields under G is uniquely determined by their transformation under the unbroken subgroup H. It is straightforward to work out the rules for using these ingredients to construct effective Lagrangians that are invariant under G as well as H.[17] Once again, the key point is that the invariance of the Lagrangian under G would eliminate all presence of the Goldstone boson field in the Lagrangian if the field were spacetime-independent, so wherever functions of this field appear in the Lagrangian they are always accompanied with at least one spacetime derivative.

[16] S. Coleman, J. Wess, and B. Zumino, *Phys. Rev.* **177**, 2239 (1969); C. G. Callan, S. Coleman, J. Wess, and B. Zumino, *Phys. Rev.* **177**, 2247 (1969).

[17] There is a complication. In some cases, such as $SU(3) \times SU(3)$ spontaneously broken to $SU(3)$, fermion loops produce G-invariant terms in the action that are not the integrals of G-invariant terms in the Lagrangian density; see J. Wess and B. Zumino, *Phys. Lett.* **37B**, 95 (1971); E. Witten, *Nucl. Phys. B* **223**, 422 (1983). The most general such terms in the action, whether or not produced by fermion loops, have been cataloged by E. D'Hoker and S. Weinberg, *Phys. Rev. D* **50**, R6050 (1994). It turns out that for $SU(N) \times SU(N)$ spontaneously broken to the diagonal $SU(N)$, there is just one such term for $N \geq 3$, and none for $N = 1$ or 2. For $N = 3$, this term is the one found by Wess and Zumino.

In the following years, effective Lagrangians with spontaneously broken $SU(2) \times SU(2)$ or $SU(3) \times SU(3)$ symmetry were widely used in lowest-order perturbation theory to make predictions about low energy pion and kaon interactions.[18] But during this period, from the late 1960s to the late 1970s, like many other particle physicists I was chiefly concerned with developing and testing the Standard Model of elementary particles. As it happened, the Standard Model did much to clarify the basis for chiral symmetry. Quantum chromodynamics with N light quarks is automatically invariant under an $SU(N) \times SU(N)$ chiral symmetry,[19] broken in the Lagrangian only by quark masses, and the electroweak theory tells us that the currents of this symmetry (along with the quark number currents) are just those to which the $W^{\pm}$, Z^0, and photon are coupled.

During this whole period, effective field theories appeared as only a device for more easily reproducing the results of current algebra. It was difficult to take them seriously as dynamical theories, because the derivative couplings that made them useful in the lowest order of perturbation theory also made them non-renormalizable, thus apparently closing off the possibility of using these theories in higher order.

My thinking about this began to change in 1976. I was invited to give a series of lectures at Erice that summer, and took the opportunity to learn the theory of critical phenomena by giving

[18]For reviews, see S. Weinberg, in *Lectures on Elementary Particles and Quantum Field Theory — 1970 Brandeis University Summer Institute in Theoretical Physics, Vol. 1*, ed. S. Deser, M. Grisaru, and H. Pendleton (The M.I.T. Press, Cambridge, MA, 1970); B. W. Lee, *Chiral Dynamics* (Gordon and Breach, New York, 1972).

[19]For a while it was not clear why there was not also a chiral $U(1)$ symmetry, that would also be broken in the Lagrangian only by the quark masses, and would either lead to a parity doubling of observed hadrons, or to a new light pseudoscalar neutral meson, both of which possibilities were experimentally ruled out. It was not until 1976 that 't Hooft pointed out that the effect of triangle anomalies in the presence of instantons produced an intrinsic violation of this unwanted chiral $U(1)$ symmetry; see G. 't Hooft, *Phys. Rev. D* **14**, 3432 (1976).

lectures about it.[20] In preparing these lectures, I was struck by Kenneth Wilson's device of "integrating out" short-distance degrees of freedom by introducing a variable ultraviolet cutoff, with the bare couplings given a cutoff dependence that guaranteed that physical quantities are cutoff independent. Even if the underlying theory is renormalizable, once a finite cutoff is introduced it becomes necessary to introduce every possible interaction, renormalizable or not, to keep physics strictly cutoff independent. From this point of view, it doesn't make much difference whether the underlying theory is renormalizable or not. Indeed, I realized that even without a cutoff, as long as every term allowed by symmetries is included in the Lagrangian, there will always be a counterterm available to absorb every possible ultraviolet divergence by renormalization of the corresponding coupling constant. Non-renormalizable theories, I realized, are just as renormalizable as renormalizable theories.

This opened the door to the consideration of a Lagrangian containing terms like (5) as the basis for a legitimate dynamical theory, not limited to the tree approximation, provided one adds every one of the infinite number of other, higher-derivative, terms allowed by chiral symmetry.[21] But for this to be useful, it is necessary that in some sort of perturbative expansion, only a finite number of terms in the Lagrangian can appear in each order of perturbation theory.

In chiral dynamics, this perturbation theory is provided by an expansion in powers of small momenta and pion masses. At momenta of order m_π, the number ν of factors of momenta or m_π contributed by a diagram with L loops, E_N external nucleon lines, and V_i vertices of type i, for any reaction among pions and/or nucleons, is

$$\nu = \sum_i V_i \left(d_i + \frac{n_i}{2} + m_i - 2\right) + 2L + 2 - \frac{E_N}{2}, \qquad (12)$$

[20]S. Weinberg, "Critical Phenomena for Field Theorists," in *Understanding the Fundamental Constituents of Matter*, ed. A. Zichichi (Plenum Press, New York, 1977).

[21]I thought it appropriate to publish this in a festschrift for Julian Schwinger; see footnote 1.

where d_i, n_i, and m_i are respectively the numbers of derivatives, factors of nucleon fields, and factors of pion mass (or more precisely, half the number of factors of u and d quark masses) associated with vertices of type i. As a consequence of chiral symmetry, the minimum possible value of $d_i + n_i/2 + m_i$ is 2, so the leading diagrams for small momenta are those with $L = 0$ and any number of interactions with $d_i + n_i/2 + m_i = 2$, which are the ones given in Eqs. (5) and (6). To next order in momenta, we may include tree graphs with any number of vertices with $d_i + n_i/2 + m_i = 2$ and just one vertex with $d_i + n_i/2 + m_i = 3$ (such as the so-called σ-term). To next order, we include any number of vertices with $d_i + n_i/2 + m_i = 2$, plus either a single loop, or a single vertex with $d_i + n_i/2 + m_i = 4$ which provides a counterterm for the infinity in the loop graph, or two vertices with $d_i + n_i/2 + m_i = 3$. And so on. Thus one can generate a power series in momenta and m_π, in which only a few new constants need to be introduced at each new order. As an explicit example of this procedure, I calculated the one-loop corrections to pion–pion scattering in the limit of zero pion mass, and of course I found the sort of corrections required to this order by unitarity.[22]

But even if this procedure gives well-defined finite results, how do we know they are true? It would be extraordinarily difficult to justify any calculation involving loop graphs using current algebra. For me in 1979, the answer involved a radical reconsideration of the nature of quantum field theory. From its beginning in the late 1920s, quantum field theory had been regarded as the application of quantum mechanics to fields that are among the fundamental constituents of the universe — first the electromagnetic field, and later the electron field and fields for other known "elementary" particles. In fact, this became a working definition of an elementary particle — it is a particle with its own field. But for years in teaching courses on quantum field theory I had emphasized that the description of nature by quantum field theories is inevitable,

---

[22]Unitarity corrections to soft-pion results of current algebra had been considered earlier by L.-F. Li and H. Pagels, *Phys. Rev. Lett.* **26**, 1204 (1971); *Phys. Rev. D* **5**, 1509 (1972); P. Langacker and H. Pagels, *Phys. Rev. D* **8**, 4595 (1973).

at least in theories with a finite number of particle types, once one assumes the principles of relativity and quantum mechanics, plus the cluster decomposition principle, which requires that distant experiments have uncorrelated results. So I began to think that although specific quantum field theories may have a content that goes beyond these general principles, quantum field theory itself does not. I offered this in my 1979 paper as what Arthur Wightman would call a folk theorem: "if one writes down the most general possible Lagrangian, including *all* terms consistent with assumed symmetry principles, and then calculates matrix elements with this Lagrangian to any given order of perturbation theory, the result will simply be the most general possible S-matrix consistent with perturbative unitarity, analyticity, cluster decomposition, and the assumed symmetry properties." So current algebra wasn't needed.

There was an interesting irony in this. I had been at Berkeley from 1959 to 1966, when Geoffrey Chew and his collaborators were elaborating a program for calculating S-matrix elements for strong interaction processes by the use of unitarity, analyticity, and Lorentz invariance, without reference to quantum field theory. I found it an attractive philosophy, because it relied only on a minimum of principles, all well established. Unfortunately, the S-matrix theorists were never able to develop a reliable method of calculation, so I worked instead on other things, including current algebra. Now in 1979 I realized that the assumptions of S-matrix theory, supplemented by chiral invariance, were indeed all that are needed at low energy, but the most convenient way of implementing these assumptions in actual calculations was by good old quantum field theory, which the S-matrix theorists had hoped to supplant.

After 1979, effective field theories were applied to strong interactions in work by Gasser, Leutwyler, Meissner, and many other theorists. My own contributions to this work were limited to two areas — isospin violation and nuclear forces.

At first in the development of chiral dynamics there had been a tacit assumption that isotopic spin symmetry was a better approximate symmetry than chiral $SU(2)\times SU(2)$, and that the Gell-Mann–Ne'eman $SU(3)$ symmetry was a better approximate symmetry than chiral $SU(3) \times SU(3)$. This assumption became untenable with the calculation of quark mass ratios from the measured pseudoscalar meson masses.[23] It turns out that the d quark mass is almost twice the u quark mass, and the s quark mass is very much larger than either. As a consequence of the inequality of d and u quark masses, chiral $SU(2)\times SU(2)$ is broken in the Lagrangian of quantum chromodynamics not only by the fourth component of a chiral four-vector, as in (6), but also by the third component of a different chiral four-vector proportional to $m_u - m_d$ (whose fourth component is a pseudoscalar). There is no function of the pion field alone, without derivatives, with the latter transformation property, which is why pion–pion scattering and the pion masses are described by (6) and the first term in (5) in leading order, with no isospin breaking aside of course from that due to electromagnetism. But there are non-derivative corrections to pion–nucleon interactions,[24] which at momenta of order m_π are suppressed relative to the derivative coupling terms in (5) by just one factor of m_π or momenta:

$$\begin{aligned}\Delta'\mathcal{L}_{\text{eff}} = &-\frac{A}{2}\left(\frac{1-\pi^2/F_\pi^2}{1+\pi^2/F_\pi^2}\right)\bar{N}N \\ &- B\left[\bar{N}t_3N - \frac{2}{F_\pi^2}\left(\frac{\pi_3}{1+\pi^2/F_\pi^2}\right)\bar{N}\vec{t}\cdot\vec{\pi}N\right] \\ &- \frac{iC}{1+\vec{\pi}^2/F_\pi^2}\bar{N}\gamma_5\vec{\pi}\cdot\vec{t}N \\ &- \frac{iD\pi_3}{1+\vec{\pi}^2/F_\pi^2}\bar{N}\gamma_5 N,\end{aligned} \tag{13}$$

where A and C are proportional to $m_u + m_d$, and B and D are proportional to $m_u - m_d$, with $B \simeq -2.5$ MeV. The A and B terms

[23] S. Weinberg, contribution to a festschrift for I. I. Rabi, *Trans. N. Y. Acad. Sci.* **38**, 185 (1977).

[24] S. Weinberg, in *Chiral Dynamics: Theory and Experiment — Proceedings of the Workshop Held at MIT, July 1994* (Springer-Verlag, Berlin, 1995). The terms in Eq. (13) that are odd in the pion field are given in Section 19.5 of S. Weinberg, *The Quantum Theory of Fields*, Vol. II (Cambridge University Press, 1996).

contribute isospin conserving and violating terms to the so-called σ-term in pion–nucleon scattering.

My work on nuclear forces began one day in 1990 while I was lecturing to a graduate class at Texas. I derived Eq. (12) for the class, and showed how the interactions in the leading tree graphs with $d_i + n_i/2 + m_i = 2$ were just those given here in Eqs. (5) and (6). Then, while I was standing at the blackboard, it suddenly occurred to me that there was one other term with $d_i+n_i/2+m_i = 2$ that I had never previously considered: an interaction with no factors of pion mass and no derivatives (and hence, according to chiral symmetry, no pions), but *four* nucleon fields — that is, a sum of Fermi interactions $(\bar{N}\Gamma N)(\bar{N}\Gamma' N)$, with any matrices Γ and Γ' allowed by Lorentz invariance, parity conservation, and isospin conservation. This is just the sort of "hard core" nucleon–nucleon interaction that nuclear theorists had long known has to be added to the pion-exchange term in theories of nuclear force. But there is a complication — in graphs for nucleon–nucleon scattering at low energy, two-nucleon intermediate states make a large contribution that invalidates the sort of power-counting that justifies the use of the effective Lagrangian (5), (6) in processes involving only pions, or one low-energy nucleon plus pions. So it is necessary to apply the effective Lagrangian, including the terms $(\bar{N}\Gamma N)(\bar{N}\Gamma' N)$ along with the terms (5) and (6), to the two-nucleon irreducible nucleon–nucleon potential, rather than directly to the scattering amplitude.[25] This program was initially carried further by Ordoñez, van Kolck, Friar, and their collaborators,[26] and eventually by several others.

The advent of effective field theories generated changes in point of view and suggested new techniques of calculation that propagated out to numerous areas of physics, some quite far removed from

[25] S. Weinberg, *Phys. Lett. B* **251**, 288 (1990); *Nucl. Phys. B* **363**, 3 (1991); *Phys. Lett. B* **295**, 114 (1992).

[26] C. Ordoñez and U. van Kolck, *Phys. Lett. B* **291**, 459 (1992); C. Ordoñez. L. Ray, and U. van Kolck, *Phys. Rev. Lett.* **72**, 1982 (1994); U. van Kolck, *Phys. Rev. C* **49**, 2932 (1994); U. van Kolck, J. Friar, and T. Goldman, *Phys. Lett. B* **371**, 169 (1996); C. Ordoñez, L. Ray, and U. van Kolck, *Phys. Rev. C* **53**, 2086 (1996); C. J. Friar, *Few-Body Systems Suppl.* **99**, 1 (1996).

particle physics. Notable here is the use of the power-counting arguments of effective field theory to justify the approximations made in the BCS theory of superconductivity.[27] Instead of counting powers of small momenta, one must count powers of the departures of momenta from the Fermi surface. Also, general features of theories of inflation have been clarified by re-casting these theories as effective field theories of the inflaton and gravitational fields.[28]

Perhaps the most important lesson from chiral dynamics was that we should keep an open mind about renormalizability. The renormalizable Standard Model of elementary particles may itself be just the first term in an effective field theory that contains every possible interaction allowed by Lorentz invariance and the $SU(3) \times SU(2) \times U(1)$ gauge symmetry, only with the non-renormalizable terms suppressed by negative powers of some very large mass M, just as the terms in chiral dynamics with more derivatives than in Eq. (5) are suppressed by negative powers of $2\pi F_\pi \approx m_N$. One indication that there is a large mass scale in some theory underlying the Standard Model is the well-known fact that the three (suitably normalized) running gauge couplings of $SU(3)\times SU(2)\times U(1)$ become equal at an energy of the order of 10^{15} GeV (or, if supersymmetry is assumed, 2×10^{16} GeV, with better convergence of the couplings.)

In 1979 both Frank Wilczek[29] and I[30] independently pointed out that, while the renormalizable terms of the Standard Model cannot violate baryon or lepton conservation,[31] this is not true

---

[27]G. Benfatto and G. Gallavotti, *J. Stat. Phys.* **59**, 541 (1990); *Phys. Rev.* **42**, 9967 (1990); J. Feldman and E. Trubowitz, *Helv. Phys. Acta* **63**, 157 (1990); **64**, 213 (1991); **65**, 679 (1992); R. Shankar, *Physica A* **177**, 530 (1991); *Rev. Mod. Phys.* **66**, 129 (1993); J. Polchinski, in *Recent Developments in Particle Theory, Proceedings of the 1992 TASI*, eds. J. Harvey and J. Polchinski (World Scientific, Singapore, 1993); S. Weinberg, *Nucl. Phys. B* **413**, 567 (1994).

[28]C. Cheung, P. Creminilli, A. L. Fitzpatrick, J. Kaplan, and L. Senatore, *JHEP* **0803**, 014 (2008); S. Weinberg, *Phys. Rev. D* **73**, 123541 (2008).

[29]F. Wilczek, *Phys. Rev. Lett.* **43**, 1570 (1979).

[30]S. Weinberg, *Phys. Rev. Lett.* **43**, 1566 (1979).

[31]This is not true if the effective theory contains fields for the squarks and sleptons of supersymmetry. However, there are no renormalizable baryon or lepton violating terms in "split supersymmetry" theories, in which the squarks and

of the higher non-renormalizable terms. In particular, four-fermion terms can generate a proton decay into antileptons, though not into leptons, with an amplitude suppressed on dimensional grounds by a factor M^{-2}. The conservation of baryon and lepton number in observed physical processes thus may be an accident, an artifact of the necessary simplicity of the leading renormalizable $SU(3) \times SU(2) \times U(1)$-invariant interactions. I also noted at the same time that interactions between a pair of lepton doublets and a pair of scalar doublets can generate a neutrino mass, which is suppressed only by a factor M^{-1}, and that therefore with a reasonable estimate of M could produce observable neutrino oscillations. The subsequent confirmation of neutrino oscillations lends support to the view of the Standard Model as an effective field theory, with M somewhere in the neighborhood of 10^{16} GeV.

Of course, these non-renormalizable terms can be (and in fact, had been) generated in various renormalizable grand unified theories by integrating out the heavy particles in these theories. Some calculations in the resulting theories can be assisted by treating them as effective field theories.[32] But the important point is that the existence of suppressed baryon- and lepton-nonconserving terms, and some of their detailed properties, should be expected on much more general grounds, *even if the underlying theory is not a quantum field theory at all.* Indeed, from the 1980s on, it has been increasingly popular to suppose that the theory underlying the Standard Model as well as general relativity is a string theory.

---

sleptons are superheavy, and only the gauginos and perhaps higgsinos survive to ordinary energies; see N. Arkani-Hamed and S. Dimopoulos, *JHEP* **0506**, 073 (2005); G. F. Giudice and A. Romanino, *Nucl. Phys. B* **699**, 65 (2004); N. Arkani-Hamed, S. Dimopoulos, G. F. Giudice, and A. Romanino, *Nucl. Phys. B* **709**, 3 (2005); A. Delgado and G. F. Giudice, *Phys. Lett. B* **627**, 155 (2005).

[32]The effective field theories derived by integrating out heavy particles had been considered by T. Appelquist and J. Carrazone, *Phys. Rev. D* **11**, 2856 (1975). In 1980, in a paper titled "Effective Gauge Theories," I used the techniques of effective field theory to evaluate the effects of integrating out the heavy gauge bosons in grand unified theories on the initial conditions for the running of the gauge couplings down to accessible energies: S. Weinberg, *Phys. Lett.* **91B**, 51 (1980).

Which brings me to gravitation. Just as we have learned to live with the fact that there is no renormalizable theory of pion fields that is invariant under the chiral transformation (7), so also we should not despair of applying quantum field theory to gravitation just because there is no renormalizable theory of the metric tensor that is invariant under general coordinate transformations. It increasingly seems apparent that the Einstein–Hilbert Lagrangian $\sqrt{g}R$ is just the least suppressed term in the Lagrangian of an effective field theory containing every possible generally covariant function of the metric and its derivatives. The application of this point of view to long range properties of gravitation has been most thoroughly developed by John Donoghue and his collaborators.[33] One consequence of viewing the Einstein–Hilbert Lagrangian as one term in an effective field theory is that any theorem based on conventional general relativity, which declares that under certain initial conditions future singularities are inevitable, must be reinterpreted to mean that under these conditions higher terms in the effective action become important.

Of course, there is a problem — the effective theory of gravitation cannot be used at very high energies, say of the order of the Planck mass, no more than chiral dynamics can be used above a momentum of order $2\pi F_\pi \approx 1\,\text{GeV}$. For purposes of the subsequent discussion, it is useful to express this problem in terms of the Wilsonian renormalization group. The effective action for gravitation takes the form

$$I_{\text{eff}} = -\int d^4x\sqrt{-\text{Det}g}\big[f_0(\Lambda) + f_1(\Lambda)R + f_{2a}(\Lambda)R^2 + f_{2b}(\Lambda)R^{\mu\nu}R_{\mu\nu} + f_{3a}(\Lambda)R^3 + \ldots\big], \tag{14}$$

where here Λ is the ultraviolet cutoff, and the $f_n(\Lambda)$ are coupling parameters with a cutoff dependence chosen so that physical

[33] J. F. Donoghue, *Phys. Rev. D* 50, 3874 (1884); *Phys. Lett.* **72**, 2996 (1994); lectures presented at the Advanced School on Effective Field Theories (Almunecar, Spain, June 1995), arXiv:gr-qc/9512024; J. F. Donoghue, B. R. Holstein, B. Garbrecth, and T. Konstandin, *Phys. Lett. B* **529**, 132 (2002); N. E. J. Bjerrum-Bohr, J. F. Donoghue, and B. R. Holstein, *Phys. Rev. D* **68**, 084005 (2003).

quantities are cutoff-independent. We can replace these coupling parameters with dimensionless parameters $g_n(\Lambda)$:

$$g_0 \equiv \Lambda^{-4} f_0; \quad g_1 \equiv \Lambda^{-2} f_1; \quad g_{2a} \equiv f_{2a};$$
$$g_{2b} \equiv f_{2b}; \quad g_{3a} \equiv \Lambda^2 f_{3a}; \ \ldots \tag{15}$$

Because dimensionless, these parameters must satisfy a renormalization group equation of the form

$$\Lambda \frac{d}{d\Lambda} g_n(\Lambda) = \beta_n(g(\Lambda)). \tag{16}$$

In perturbation theory, all but a finite number of the $g_n(\Lambda)$ go to infinity as $\Lambda \to \infty$, which if true would rule out the use of this theory to calculate anything at very high energy. There are even examples, like the Landau pole in quantum electrodynamics and the phenomenon of "triviality" in scalar field theories, in which the couplings blow up at a *finite* value of Λ.

It is usually assumed that this explosion of the dimensionless couplings at high energy is irrelevant in the theory of gravitation, just as it is irrelevant in chiral dynamics. In chiral dynamics, it is understood that at energies of order $2\pi F_\pi \approx m_N$, the appropriate degrees of freedom are no longer pion and nucleon fields, but rather quark and gluon fields. In the same way, it is usually assumed that in the quantum theory of gravitation, when Λ reaches some very high energy, of the order of 10^{15} to 10^{18} GeV, the appropriate degrees of freedom are no longer the metric and the Standard Model fields, but something very different, perhaps strings.

But maybe not. It is just possible that the appropriate degrees of freedom at all energies are the metric and matter fields, including those of the Standard Model. The dimensionless couplings can be protected from blowing up if they are attracted to a finite value g_{n*}. This is known as *asymptotic safety.*[34]

Quantum chromodynamics provides an example of asymptotic safety, but one in which the theory at high energies is not only safe from exploding couplings, but also free. In the more general

[34]This was first proposed in my 1976 Erice lectures; see footnote 20.

case of asymptotic safety, the high energy limit g_{n*} is finite, but not commonly zero.

For asymptotic safety to be possible, it is necessary that all the beta functions should vanish at g_{n*}:

$$\beta_n(g_*) = 0. \tag{17}$$

It is also necessary that the physical couplings should be on a trajectory that is attracted to g_{n*}. The number of independent parameters in such a theory equals the dimensionality of the surface, known as the *ultraviolet critical surface*, formed by all the trajectories that are attracted to the fixed point. This dimensionality had better be finite, if the theory is to have any predictive power at high energy. For an asymptotically safe theory with a finite-dimensional ultraviolet critical surface, the requirement that couplings lie on this surface plays much the same role as the requirement of renormalizability in the quantum chromodynamics — it provides a rational basis for limiting the complexity of the theory.

This dimensionality of the ultraviolet critical surface can be expressed in terms of the behavior of $\beta_n(g)$ for g near the fixed point g_*. Barring unexpected singularities, in this case we have

$$\beta_n(g) \to \sum_m B_{nm}(g_m - g_{*m}), \quad B_{nm} \equiv \left(\frac{\partial \beta_n(g)}{\partial g_m}\right)_* . \tag{18}$$

The solution of Eq. (16) for g near g_* is then

$$g_n(\Lambda) \to g_{n*} + \sum_i u_{in}\, \Lambda^{\lambda_i}, \tag{19}$$

where λ_i and u_{in} are the eigenvalues and suitably normalized eigenvectors of B_{nm}:

$$\sum_m B_{nm}\, u_{im} = \lambda_i\, u_{in}. \tag{20}$$

Because B_{nm} is real but not symmetric, the eigenvalues are either real, or come in pairs of complex conjugates. The dimensionality of the ultraviolet critical surface is therefore equal to the number of eigenvalues of B_{nm} with negative real part. The condition that the couplings lie on this surface can be regarded as a generalization of the condition that quantum chromodynamics, if it were a fundamental and not merely an effective field theory, would have to involve only renormalizable couplings.

It may seem unlikely that an infinite matrix like B_{nm} should have only a finite number of eigenvalues with negative real part, but in fact examples of this are quite common. As we learned from the Wilson–Fisher theory of critical phenomena, when a substance undergoes a second-order phase transition, its parameters are subject to a renormalization group equation that has a fixed point, with a single infrared-repulsive direction, so that adjustment of a single parameter such as the temperature or the pressure can put the parameters of the theory on an infrared attractive surface of co-dimension one, leading to long-range correlations. The single infrared-repulsive direction is at the same time a unique ultraviolet-attractive direction, so the ultraviolet critical surface in such a theory is a one-dimensional curve. Of course, the parameters of the substance on this curve do not really approach a fixed point at very short distances, because at a distance of the order of the interparticle spacing the effective field theory describing the phase transition breaks down.

What about gravitation? There are indications that here too there is a fixed point, with an ultraviolet critical surface of finite dimensionality. Fixed points have been found (of course with $g_{n*} \neq 0$) using dimensional continuation from $2 + \epsilon$ to 4 spacetime dimensions,[35] by a $1/N$ approximation (where N is the number of added matter fields),[36] by lattice methods,[37] and by use of the truncated

[35] S. Weinberg, in *General Relativity*, ed. S. W. Hawking and W. Israel (Cambridge University Press, 1979): 700; H. Kawai, Y. Kitazawa, & M. Ninomiya, *Nucl. Phys. B* **404**, 684 (1993); *Nucl. Phys. B* **467**, 313 (1996); T. Aida & Y. Kitazawa, *Nucl. Phys. B* **401**, 427 (1997); M. Niedermaier, *Nucl. Phys. B* **673**, 131 (2003).

[36] L. Smolin, *Nucl. Phys. B* **208**, 439 (1982); R. Percacci, *Phys. Rev. D* **73**, 041501 (2006).

[37] J. Ambjørn, J. Jurkewicz, & R. Loll, *Phys. Rev. Lett.* **93**, 131301 (2004); *Phys. Rev. Lett.* **95**, 171301 (2005); *Phys. Rev. D* **72**, 064014 (2005); *Phys. Rev. D* **78**, 063544 (2008); and in *Approaches to Quantum Gravity*, ed. D. Oríti (Cambridge University Press).

exact renormalization group equation,[38] initiated in 1968 by Martin Reuter. In the last method, which had earlier been introduced in condensed matter physics[39] and then carried over to particle theory,[40] one derives an exact renormalization group equation for the total vacuum amplitude $\Gamma[g, \Lambda]$ in the presence of a background metric $g_{\mu\nu}$ with an *infrared* cutoff Λ. This is the action to be used in calculations of the true vacuum amplitude in calculations of graphs with an *ultraviolet* cutoff Λ. To have equations that can be solved, it is necessary to truncate these renormalization group equations, writing $\Gamma[g, \Lambda]$ as a sum of just a finite number of terms like those shown explicitly in Eq. (14), and ignoring the fact that the beta function inevitably does not vanish for the couplings of other terms in $\Gamma[g, \Lambda]$ that in the given truncation are assumed to vanish.

Initially only two terms were included in the truncation of $\Gamma[g, \Lambda]$ (a cosmological constant and the Einstein–Hilbert term $\sqrt{g}R$), and a fixed point was found with two eigenvalues λ_i, a pair of complex conjugates with negative real part. Then a third operator ($R_{\mu\nu}R^{\mu\nu}$ or the equivalent) was added, and a third eigenvalue was found, with λ_i real and negative. This was not encouraging. If each time that

[38] M. Reuter, *Phys. Rev. D* **57**, 971 (1998); D. Dou & R. Percacci, *Class. Quant. Grav.* **15**, 3449 (1998); W. Souma, *Prog. Theor. Phys.* **102**, 181 (1999); O. Lauscher & M. Reuter, *Phys. Rev. D* **65**, 025013 (2001); *Class. Quant. Grav.* **19**, 483 (2002); M. Reuter & F. Saueressig, *Phys. Rev. D* **65**, 065016 (2002); O. Lauscher & M. Reuter, *Int. J. Mod. Phys. A* **17**, 993 (2002); *Phys. Rev. D* **66**, 025026 (2002); M. Reuter and F. Saueressig, *Phys Rev. D* **66**, 125001 (2002); R. Percacci & D. Perini, *Phys. Rev. D* **67**, 081503 (2002); *Phys. Rev. D* **68**, 044018 (2003); D. Perini, *Nucl. Phys. Proc. Suppl. C* **127**, 185 (2004); D. F. Litim, *Phys. Rev. Lett.* **92**, 201301 (2004); A. Codello & R. Percacci, *Phys. Rev. Lett.* **97**, 221301 (2006); A. Codello, R. Percacci, & C. Rahmede, *Int. J. Mod. Phys. A* **23**, 143 (2008); M. Reuter & F. Saueressig, arXiv:0708.1317; P. F. Machado and F. Saueressig, *Phys. Rev. D* **77**, 124045 (2008); A. Codello, R. Percacci, & C. Rahmede, *Ann. Phys.* **324**, 414 (2009); A. Codello & R. Percacci, arXiv:0810.0715; D. F. Litim arXiv:0810.3675; H. Gies & M. M. Scherer, arXiv:0901.2459; D. Benedetti, P. F. Machado, & F. Saueressig, arXiv:0901.2984, 0902.4630; M. Reuter & H. Weyer, arXiv:0903.2971.

[39] F. J. Wegner and A. Houghton, *Phys. Rev. A* **8**, 401 (1973).

[40] J. Polchinski, *Nucl. Phys. B* **231**, 269 (1984); C. Wetterich, *Phys. Lett. B* **301**, 90 (1993).

new terms were included in the truncation, new eigenvalues appeared with negative real part, then the ultraviolet critical surface would be infinite dimensional, and the theory, though free of couplings that exploded at high energy, would lose all predictive value at high energy.

In just the last few years calculations have been done that allow more optimism. Codello, Percacci, and Rahmede[41] have considered a Lagrangian containing all terms $\sqrt{g}R^n$ with n running from zero to a maximum value $n_{\rm max}$, and find that the ultraviolet critical surface has dimensionality 3 even when $n_{\rm max}$ exceeds 2, up to the highest value $n_{\rm max} = 6$ that they considered, for which the space of coupling constants is 7-dimensional. Furthermore, the three eigenvalues they find with negative real part seem to converge as $n_{\rm max}$ increases, as shown in the following table of ultraviolet-attractive eigenvalues:

$n_{\rm max} = 2:$	$-1.38 \pm 2.32i$	-26.8
$n_{\rm max} = 3:$	$-2.71 \pm 2.27i$	-2.07
$n_{\rm max} = 4:$	$-2.86 \pm 2.45i$	-1.55
$n_{\rm max} = 5:$	$-2.53 \pm 2.69i$	-1.78
$n_{\rm max} = 6:$	$-2.41 \pm 2.42i$	-1.50

In a subsequent paper[42] they added matter fields, and again found just three ultraviolet-attractive eigenvalues. Further, this year Benedetti, Machado, and Saueressig[43] considered a truncation with a different four terms, terms proportional to $\sqrt{g}R^n$ with $n = 0, 1$ and 2 and also $\sqrt{g}C_{\mu\nu\rho\sigma}C^{\mu\nu\rho\sigma}$ (where $C_{\mu\nu\rho\sigma}$ is the Weyl tensor) and they too find just three ultraviolet-attractive eigenvalues, also when matter is added. If this pattern of eigenvalues continues to hold in future calculations, it will begin to look as if there is a quantum field theory of gravitation that is well-defined at all energies, and that has just three free parameters.

The natural arena for application of these ideas is in the physics of gravitation at small distance scales and high energy — specifically, in

[41] A. Codello, R. Percacci, & C. Rahmede, Int. J. Mod. Phys. A23, 143 (2008).
[42] A. Codello, R. Percacci, & C. Rahmede, *Ann. Phys.* **324**, 414 (2009).
[43] D. Benedetti, P. F. Machado, & F. Saueressig, arXiv:0901.2984, 0902.4630.

the early universe. A start in this direction has been made by Reuter and his collaborators,[44] but much remains to be done.

I am grateful for correspondence about recent work on asymptotic safety with D. Benedetti, D. Litim, R. Percacci, and M. Reuter, and to G. Colangelo and J. Gasser for inviting me to give this talk.

[44] A. Bonanno and M. Reuter, *Phys. Rev. D* **65**, 043508 (2002); *Phys. Lett. B* **527**, 9 (2002); M. Reuter and F. Saueressig, *J. Cosm. and Astropart. Phys.* **09**, 012 (2005).

Chapter 2

Yoichiro Nambu and the origin of mass

Tom W.B. Kibble

Blackett Laboratory, Imperial College London
SW7 2AZ, UK

This note summarizes and celebrates the important contributions of Yoichiro Nambu to the tricky question of the origin of particle masses.

The origin of the masses of elementary particles has been a major puzzle since the early days of particle physics, and remains so today. For example, we have no real idea of where the quark masses come from. There has however been important progress, including the very significant development of the idea of spontaneous symmetry breaking in gauge theories. In that development many of the key early steps were taken by Yoichiro Nambu. In the excitement that followed the very probable discovery in 2012 of the Higgs boson at the LHC in CERN, most of the attention was on later theoretical developments, in particular the so-called Higgs mechanism, and perhaps this important early work was unfairly neglected.

1. Physics after the Second World War

There was a great flowering of particle physics in the years following the end of the Second World War, when large numbers of physicists who had been working on military projects returned to their university laboratories.

On the theoretical side the great triumph was the development of renormalization theory independently in 1947 by Julian Schwinger [1] and by Richard Feynman [2], and earlier, in 1943, by Sin-ichiro Tomonaga [3]. This showed that calculations beyond the lowest order in the fine structure constant are possible in quantum electrodynamics (QED). At the same time, on the experimental side the beautiful experiments of Willis Lamb [4] provided an extremely accurate value for the Lamb shift in hydrogen, the separation between the $2s_{1/2}$ and $2p_{1/2}$ energy levels. The theoretical calculations were able to fit that result and the measured anomalous magnetic moment of the electron with unprecedented accuracy.

This triumphant success of QED encouraged particle physicists to search for similarly successful theories of the other forces, the strong and weak nuclear interactions. Since QED is a gauge theory, and gauge invariance plays an important role in rendering the theory consistent (via the Ward identities), it was natural to ask whether the other forces could also be described by gauge theories. The first suggested gauge theory beyond QED was that of Yang and Mills in 1954 [5], intended as a theory of strong interactions based on a gauged version of the SU(2) isospin symmetry. The same theory was in fact written down in the same year by Ronald Shaw, a student of Abdus Salam's in Cambridge, although it was never published except as part of a Cambridge University PhD thesis [6].

There was however an immediately obvious problem with this or any gauge theory of the strong or weak interactions. How could we understand the short range of these forces, demanding massive intermediate force carriers, when the gauge bosons are apparently naturally massless? Indeed it was widely seen as one of the successes of QED that gauge invariance correctly predicted the masslessness of the photon. The issue was somewhat clouded in the case of the Yang–Mills theory by the fact that isospin symmetry is only approximate. But this still left a big puzzle. It was well known that simply adding a mass term for the vector field into the Lagrangian would yield a non-renormalizable, and hence inconsistent, theory. So how could the theory consistently accommodate massive vector particles?

The first person to note that gauge invariance does *not* in fact require the photon to be massless was Julian Schwinger in 1957 [7]. If the photon's interaction were strong enough it could in fact acquire a non-zero mass.

2. Spontaneous symmetry breaking in superconductivity

Outside of particle physics, the great advance of the nineteen-fifties was the work of Bardeen, Cooper and Schrieffer [8] that provided the first real understanding of the mechanism of superconductivity. The phonon-mediated interaction can bind pairs of electrons with opposite momenta and spins to form *Cooper pairs.* When the system is cooled below the critical temperature, these form a Bose–Einstein condensate, with macroscopic occupation of a single quantum state.

Prior to that development, a very useful phenomenological model of a superconductor was provided by Ginzburg and Landau [9], with a scalar order-parameter field $\phi(t, \boldsymbol{x})$ representing the Cooper pairs. Here $|\phi^2|$ represents the number density of pairs. Since the pairs are formed of two electrons, the appropriate gauge transformation is

$$\phi \to \phi e^{2ie\lambda}, \quad \boldsymbol{A} \to \boldsymbol{A} + \boldsymbol{\nabla}\lambda. \tag{1}$$

The Hamiltonian (with $\hbar = c = 1$) is

$$H = \int d^3\boldsymbol{x} \left[\frac{1}{2m} \boldsymbol{D}\phi^* \cdot \boldsymbol{D}\phi + V(\phi) \right], \tag{2}$$

where the covariant derivative is

$$\boldsymbol{D}\phi = \boldsymbol{\nabla}\phi - 2e\boldsymbol{A}\phi, \tag{3}$$

and the effective potential V is a function only of $\phi^*\phi$. Near to the critical temperature T_c, it may be expanded as a power series, with temperature-dependent coefficients, keeping only the lowest terms,

$$V = \alpha(T)\phi^*\phi + \tfrac{1}{2}\beta(T)(\phi^*\phi)^2. \tag{4}$$

As the temperature is lowered through T_c, α changes sign, and the bowl shape of V is replaced by sombrero shape, where the minimum energy is found not at $\phi = 0$ but around the circle

$$\phi^* \phi = \frac{-\alpha}{\beta}. \tag{5}$$

So in the minimum-energy state, ϕ acquires a non-zero expectation value, somewhere around this circle, implying a spontaneous breaking of the gauge symmetry. There is a degenerate family of ground states, labelled by the phase of the order parameter. Moreover, because the curvature of the potential in the radial direction is non-zero, an effective mass is generated for the gauge boson. The photon becomes a massive plasmon. At zero temperature, the plasmon mass $m_{\rm pl}$ is given by

$$m_{\rm pl}^2 = \frac{e^2 n_e}{\epsilon_0 m_e}. \tag{6}$$

In our units, $m_{\rm pl}$ is the same thing as the plasma frequency $\omega_{\rm pl}$, the minimum frequency with which electromagnetic waves can propagate within the superconductor.

This idea of broken gauge symmetry was quite controversial at the time, with many physicists insisting that the system must ultimately have a definite particle number and hence completely uncertain phase; the true ground state must be a gauge-invariant superposition of these degenerate states. It is now commonplace to talk of superconducting states with definite phase, but that notion took a long time to be accepted. Nambu [10] gave a particularly clear account of the situation, using a modified form of the Hartree–Fock approximation, demonstrating the intimate connection between symmetry breaking and the existence of an energy gap. In a closely related paper, Bogoliubov [11] pointed out that quasiparticles do not have a definite charge; a quasiparticle is a linear combination of an electron with momentum $\boldsymbol{p}$ and spin up, say, and a hole in the state with momentum $-\boldsymbol{p}$ and spin down.

3. The superconductor analogy

Nambu went on to suggest that elementary particle masses might be generated by a similar mechanism. He constructed a model to illustrate this possibility, though not a gauge theory. As a possible model of strong interactions, he envisaged a model based on a four-fermion interaction, with symmetries under both ordinary and chiral global transformations:

$$\psi(x) \to e^{i\alpha}\psi(x), \quad \psi(x) \to e^{\alpha\gamma_5}\psi(x), \tag{7}$$

with corresponding conserved Noether currents:

$$j^\mu = \bar{\psi}\gamma^\mu\psi, \quad j_5^\mu = \bar{\psi}i\gamma^\mu\gamma_5\psi. \tag{8}$$

By analogy with superconductivity, he suggested that the nucleon mass could be generated by spontaneous breaking of the chiral symmetry.

Nambu and Jona-Lasinio [12] then constructed a specific model to illustrate this idea, based on the Lagrangian

$$\mathcal{L} = i\bar{\psi}\gamma^\mu\partial_\mu\psi + g[(\bar{\psi}\psi)^2 - (\bar{\psi}\gamma_5\psi)^2]. \tag{9}$$

Symmetry breaking occurs when a non-zero expectation value $\langle 0|\bar{\psi}\psi|0\rangle$ is formed. They showed that this would imply a non-zero mass for the quasiparticle, here identified with the nucleon.

A key feature of the model is the appearance of a massless scalar boson. As shown by Goldstone [13], this is inevitable in a relativistic theory with a spontaneously broken global symmetry.

Of course, the Nambu–Jona-Lasinio model turned out not to be the correct description of strong or weak interactions, but it was nevertheless a very significant step on the way to developing one.

It was widely believed in the particle physics community that the same result must hold also for a local gauge theory [14], and that these massless Nambu–Goldstone bosons would necessarily appear. This seemed to present an insurmountable obstacle, since no such bosons had been seen, although they should have been easy to spot. It took a couple of years for particle physicists to understand how this could come about, via the so-called Higgs mechanism, embodied in three independent papers published in *Physical Review Letters* in 1964 [15–17].

4. Conclusions

These three papers attracted very little interest — and quite a lot of scepticism — for the first two or three years, but they laid another part of the groundwork for the development of the unified electroweak theory of Weinberg [18] and Salam [19], now an established component of the Standard Model and triumphantly vindicated by the almost certain discovery in 2012 of the Higgs boson at the Large Hadron Collider in CERN [20, 21].

It is clear that Nambu made many of the essential early contributions towards the eventual development of this theory, with unique insights into both particle physics and superconductivity. Without his seminal work, it would surely have taken much longer to arrive at that goal.

References

[1] J. S. Schwinger, On quantum electrodynamics and the magnetic moment of the electron, *Phys. Rev.* **73** (1948) 416.
[2] R. P. Feynman, A relativistic cutoff for classical electrodynamics, *Phys. Rev.* **74** (1948) 939.
[3] S. Tomonaga, On a relativistically invariant formulation of the quantum theory of wave fields, *Prog. Theor. Phys.* **1** (1946) 27.
[4] W. E. Lamb and R. C. Retherford, Fine structure of the hydrogen atom by a microwave method, *Phys. Rev.* **72** (1947) 241.
[5] C. N. Yang and R. L. Mills, Conservation of isotopic spin and isotopic gauge invariance, *Phys. Rev.* **96** (1954) 191.
[6] R. Shaw, Invariance under general isotopic gauge transformations. Cambridge University PhD thesis, Part II, Chapter III (1955), unpublished.
[7] J. S. Schwinger, A theory of the fundamental interactions, *Annals Phys.* **2** (1957) 407.
[8] J. Bardeen, L. N. Cooper and J. R. Schrieffer, Theory of superconductivity, *Phys. Rev.* **108** (1957) 1175.
[9] V. L. Ginzburg and L. D. Landau, On the theory of superconductivity, *Zh. Eksp. Teor. Fiz.* **20** (1950) 1064.
[10] Y. Nambu, *Phys. Rev.* **117** (1960) 648. doi:10.1103/PhysRev.117.648.
[11] N. N. Bogoliubov, A new method in the theory of superconductivity I, *Sov. Phys.-JETP* **7** (1958) 41–46 [*J. Exper. Theor. Phys.* (USSR) **34** (1958) 58–65].

[12] Y. Nambu and G. Jona-Lasinio, *Phys. Rev.* **122** (1961) 345. doi:10.1103/PhysRev.122.345.
[13] J. Goldstone, Field theories with superconductor solutions, *Nuovo Cim.* **19** (1961) 154.
[14] J. Goldstone, A. Salam and S. Weinberg, Broken symmetries, *Phys. Rev.* **127** (1962) 965.
[15] F. Englert and R. Brout, Broken symmetry and the mass of gauge vector bosons, *Phys. Rev. Lett.* **13** (1964) 321.
[16] P. W. Higgs, Broken symmetries and the masses of gauge bosons, *Phys. Rev. Lett.* **13** (1964) 508.
[17] G. S. Guralnik, C. R. Hagen and T. W. B. Kibble, Global conservation laws and massless particles, *Phys. Rev. Lett.* **13** (1964) 585.
[18] S. Weinberg, *Phys. Rev. Lett.* **19** (1967) 1264.
[19] A. Salam, Weak and electromagnetic interactions, in *Elementary Particle Theory: Proceedings of the Nobel Symposium*, Lerum, Sweden, 1968, N. Svartholm (ed.), (Almqvist & Wiksell, Stockholm, 1968), Conf. Proc. C, Vol. 680519, pp. 367–377.
[20] G. Aad *et al.* [ATLAS Collaboration], *Phys. Lett. B* **716** (2012) 1.
[21] S. Chatrchyan *et al.* [CMS Collaboration], *Phys. Lett. B* **716** (2012) 30.

Chapter 3

Yoichiro Nambu: Visionary theorist who shaped modern particle physics*

Michael S. Turner

Kavli Institute for Cosmological Physics
University of Chicago, Illinois, USA
mturner@kicp.uchicago.edu

Yoichiro Nambu was one of the most influential theoretical physicists of the twentieth century. His deep and unexpected insights often took years for others to understand and fully appreciate. They include: spontaneous symmetry breaking, for which he was awarded half of the 2008 Nobel Prize in Physics; the theory of quarks and gluons; and string theory.

Modern particle theory is defined by its accomplishments, largely embodied in the Standard Model of the strong, weak and electromagnetic interactions, and by its aspirations — a theory that unifies all the forces and particles. Nambu's contributions to symmetry breaking and the theory of quarks form the foundation of the Standard Model, and string theory is the most promising approach to a theory of everything.

Nambu, who died of a heart attack on 5 July in Osaka, Japan, was born in Tokyo in 1921. It was the year that Yoshio Nishina visited Copenhagen and brought back quantum mechanics

*This article was first published in *Nature* **524**, 416 (2015).

to Kyoto, Japan's first foray into modern physics. The 'Copenhagen in Kyoto' school included Hideki Yukawa, who won the 1949 Nobel Prize for his prediction of the existence of mesons, and Sin-Itiro Tomonaga, who shared the 1965 Nobel Prize for his work in quantum electrodynamics, the theory that describes all of electromagnetism from chemistry to lasers.

Nambu attended the University of Tokyo, graduating with a master's degree in physics in 1942. His studies were interrupted by the Second World War. In the army he dug trenches and worked on the Japanese radar project, but his mind was on fundamental physics. In 1945, he married his assistant, Chieko Hida.

Under difficult post-war circumstances, always hungry and living in his office at the University of Tokyo, Nambu finished his PhD in 1952. Although his department's research focus was condensed-matter physics, Nambu was drawn to nuclear and particle physics and he attended seminars on these topics by Nishina, Tomonaga and Yukawa at the nearby Tokyo University of Education.

In 1950, Tomonaga recommended Nambu for a faculty position at Osaka City University, where he wrote two remarkable papers. He derived the now-famous Bethe–Salpeter equation that describes the quantum theory of how particles bind together. [1] And he proposed how the newly discovered 'strange' particles were produced. [2] Each paper pre-dated by a year its more well-known counterpart written by US physicists.

Nambu's big break came in 1952 when, at Tomonaga's suggestion, he was invited by Robert Oppenheimer to the Institute for Advanced Study in Princeton, New Jersey. Years later he described that experience as overwhelming — he felt surrounded by people smarter and more aggressive than him. Nonetheless, physicist Murph Goldberger thought highly enough of him to invite him to the University of Chicago, Illinois, in 1954.

In particle physics, Chicago was the place to be just after the Second World War. Enrico Fermi was the intellectual leader of a physics department that included more than ten future Nobel Prize winners. Nambu spent the rest of his academic career — more than half a century — at the university's Enrico Fermi Institute.

With Giovanni Jona-Lasinio in 1961, Nambu introduced the idea of hidden or broken symmetries while trying to understand superconductivity — the resistanceless flow of electric current at very low temperatures. Mathematical symmetries in Maxwell's theory of electromagnetism are hidden at very low temperatures, as is the symmetry between the electromagnetic and weak forces, the hallmark of the electroweak theory. The Higgs boson, discovered in 2012 at CERN, Europe's particle-physics laboratory near Geneva, Switzerland, reveals the fact that the electroweak symmetry is broken.

In 1964, George Zweig and Murray Gell-Mann each independently proposed the idea of quarks to explain the hundreds of new elementary particles that were being discovered at particle accelerators. It took more than 20 years to sort out quarks' properties and how they are held together in triplets and pairs by a 'colour' force mediated by gluons to form protons, neutrons, mesons and other particles. But Nambu and Moo-Young Han put most of it together in 1965. As Gell-Mann said: "He did this [...] while the rest of us were floundering." In an attempt to better understand the colour force, Nambu went on to co-invent string theory.

I had the good fortune of being Yoichiro's colleague for more than 30 years. He was surprisingly soft-spoken and modest for someone so wise and important. We all listened carefully to anything he had to say, but rarely fully comprehended it. "People don't understand him, because he is so far-sighted," Edward Witten of the Institute for Advanced Study once said.

A downside of being so ahead of the times is that recognitions come slowly. After years of hoping, we were ecstatic when he received the Nobel Prize. Nambu could not travel to the ceremony in Stockholm, so the Swedish ambassador to the United States came to Chicago to present the modest giant of particle physics with his prize at a ceremony attended by 200 of his friends and colleagues. A more joyful event I cannot remember.

References

[1] Y. Nambu, *Prog. Theor. Phys.* **5**, 614–633; 1950.
[2] Y. Nambu *et al.*,*Prog. Theor. Phys.* **6**, 615–619; 1951.

Chapter 4

Yoichiro Nambu*

Peter G. O. Freund[†], Jeffrey Harvey[†], Emil Martinec[†], Pierre Ramond[‡]

[†] *University of Chicago, Illinois*
[‡] *University of Florida, Gainesville*

On 5 July 2015, at the age of 94, Yoichiro Nambu, one of the truly great theoretical physicists of our time, died in Osaka, Japan, due to an acute myocardial infarction.

Most of the important physics theories of the second half of the 20th century contain a seminal contribution by Yoichiro. We mention but three: spontaneous symmetry breaking, color gauging, and string theory, which all owe their existence to Yoichiro's deep insights. Indeed, when an idea is introduced in particle physics, it often turns out to have been already developed by Yoichiro years earlier. He was the recipient of virtually all major physics prizes, including the Nobel Prize in Physics in 2008 and the Wolf Prize in 1995. But such was the modesty of the man that one always thought of him as the creator of this or that fundamental theory and not as a Nobel or other prize laureate.

Yoichiro was born in Tokyo on 18 January 1921. His youth was affected by World War II. He served in the Japanese Army and was assigned to keep an eye on Sin-itiro Tomonaga, who was developing radar for one of the other military services. That assignment brought

*This article was first published in *Physics Today* **68**(10), 60 (2015).

Yoichiro in contact with Tomonaga's physics ideas, and after the war, though not a member of his group, Yoichiro kept up to date with its work. He set out on his own to calculate the electron's anomalous magnetic moment and obtained the famous $\alpha/2\pi$ correction. He was not aware of similar work elsewhere because Douglas MacArthur, to make sure the Japanese did not develop nuclear weapons of their own, forbade the import of US physics journals to Japan. Instead, he encouraged the Japanese to read Time magazine. It was in Time that Yoichiro read an article about Julian Schwinger's calculation of the electron's anomalous magnetic moment, which reached the same result as the one he had obtained and which thereby made Yoichiro's work no longer publishable.

Yoichiro Nambu

Two years before receiving his PhD from Tokyo Imperial University in 1952, Yoichiro was appointed an associate professor at Osaka City University. In that capacity he published two papers whose results are often quoted under the names of physicists who rediscovered them. In one, Yoichiro derived the quantum field theoretic bound-state equation usually known as the Bethe–Salpeter equation. In another, with Kazuhiko Nishijima and Yoshio Yamaguchi, he proposed the mechanism of associated production of strange particles a year before Abraham Pais did.

Yoichiro realized that the center of worldwide theoretical physics research was in the US, and in 1952 he went to the Institute for Advanced Study in Princeton, New Jersey. In 1954 he moved to the University of Chicago, where he would spend the rest of his career. He caught the tail end of Chicago's Enrico Fermi era.

In the mid-1950s, dispersion theory was center stage, and with Geoffrey Chew, Marvin Goldberger, and Francis Low, Yoichiro wrote the influential CGLN papers on meson scattering and photoproduction. From an analysis of the Stanford nucleon form-factor data, he predicted in 1957 the existence of the isospin-zero vector meson ω, which was confirmed in 1961.

At the University of Illinois in Urbana, the brilliant Bardeen–Cooper–Schrieffer theory of superconductivity was being developed. In an important paper, Yoichiro solved the problem of the theory's apparent lack of gauge invariance, and in the fundamental papers he wrote with Giovanni Jona-Lasinio, he transported the basic idea of that work into relativistic quantum field theory. Those contributions marked the birth of the fundamental theory of spontaneous symmetry breaking and also led to the Brout–Englert–Higgs mechanism by which gauge fields acquire mass.

Yoichiro made the fundamental observation that whereas the laws of nature exhibit all kinds of exact symmetries, the ground state — the vacuum — can violate those symmetries, and that by itself can result in all the effects we normally associate with symmetry breaking. The signature of the spontaneous symmetry breaking mechanism is the appearance of a massless particle, the so-called Nambu–Goldstone boson, which interacts in a characteristic manner with other particles.

In 1965, with Moo-Young Han, Yoichiro set up a model of strong interactions based on a gauge treatment of a color symmetry similar to that in Wally Greenberg's quark parastatistics. Both color gauging and spontaneous symmetry breaking are crucial to the Standard Model of particle physics.

Four years later Yoichiro and, independently, Holger Bech Nielsen and Leonard Susskind showed that the Veneziano four-point amplitudes and their N-point ($N > 4$) generalizations call for abandoning the picture of point-like elementary particles and replacing it by extended one-dimensional objects, strings. That work has led to a vast scientific enterprise still going strong today. Yoichiro approached physics with his characteristic deep and creative curiosity and took great pleasure in his work. His keen insights were driven by a marvelous and unique form of intuition. His reasoning was clear and convincing, but it was hard to find out how those superb ideas arose in his thinking. In a typically Japanese manner, Yoichiro was unable to use the word "no." Even if a preposterous request was made of him, he would finally "agree" to it, but the more preposterous he found the request, the longer the time he took before saying yes. A "yes" delivered after an infinite pause was his version of the word "no." That led to some funny situations while he was chair of the University of Chicago department of physics.

At a personal level, Yoichiro was a kind and understanding colleague, who established a pleasant and cordial atmosphere at the Enrico Fermi Institute. With his passing, we lose one of the few dominant figures who set the direction in which theoretical physics is evolving.

Chapter 5

Yoichiro Nambu*

Jorge Willemsen

RSMAS/AMP, University of Miami
4600 Rickenbacker Causeway
Miami, FL 33149
jwillemsen@rsmas.miami.edu

Nambu was many things to me. Teacher, mentor, friend.

First, as teacher he treated all of his graduate students in the same manner, a manner very distinct from what was commonplace among, say, experimental high energy physicists, and today I see it within my present setting. He never had an agenda, a set of tasks his students were to pursue in order to further his own funding let alone reputation. I would say the core of his teaching was "If you are going to make it you are going to have to make it on your own". That is, formulate a problem you are interested in and wish to solve, then I will help you, but I will not assign a problem to you. And indeed so many young Ph.D.s do their dissertation over and over again because they did it under strict supervision of their advisors and never had a chance to develop their own originality.

As a mentor he was equally strict. He took me and Tom Bell to SLAC one year when he took a sabbatical there. I loved the place and especially the people there. I desperately wanted to get a Post-Doc. It did not happen immediately. Far into the process I went to see

*This article was first published in *Physics Today Online*.

Sid Drell to ask if I still had a chance. He told me to go see Nambu and ask him to go see him. I told Nambu this and he said, "I have written my letter. I will do no more." This was, I believe, a tribute to his credit, to his nobility if you will, that he did not ever want to be a part of the "Ol'e boys club." Once again, I had to stand on my own. I got the offer and that forever shaped my future.

As to that future, there was a festival at University of Chicago to honor his birthday (65th I believe). After formal talks etc. there was a formal dinner and guests were asked to offer tributes if they so wished. I don't remember my exact words but the essence of what I said was something like "When I first went to work with Nambu I thought maybe I had caught him past his prime. He was thinking about things no one else was thinking about, talking about things no one else was talking about. Maybe this was a big mistake on my part." Amazingly he shook his head as though perhaps in agreement. But then I noted "that was just before he issued his theory of strings and all of a sudden I realized that I was in the presence of sheer genius." And I admitted openly "I know I will never rise to his level, but I will do what I can and always know that to have been in his presence enriched my life in a manner that I could not ask for more."

Finally, as friend, it turned out that in the midst of my joy at SLAC my son developed what was then a 50/50 chance of survival disease. Nambu learned of this and then shared with me the pain he had endured when his son Albert had developed a similar life-threatening disease. There was no need for this save his warmth and openness to his humble student.

Years later I was invited together with a group from my present institution to visit Japan to try to set up a research collaboration. I knew he was in Japan at the time so I contacted him and he told me that fortuitously he would be in Tokyo for a meeting just around the time of my trip and so suggested I come a few days early, he would meet me at the airport. The flight was about 6 hours late but there he was, waiting for me. He had set up for us to take the bullet train with a stop at Kyoto, which he insisted I must visit during my stay in Japan. There is a comical aside in that I lugged my bag up a set of stairs only to find that it was the wrong landing, so down we went again and over to where the bullet train was almost ready to depart. We made it.

Once I departed the train in Kyoto he told me to walk down the stairs, take my first right corner, walk a block, then take the next right as well, and there would be the hotel where he had made a reservation for me. The staff spoke no English but there they were, waiting for me because Nambu had alerted them to my arrival.

That was a few years before he was awarded the Nobel. I think, how many men of that calibre care so much about their humble students?

I can only hope he now knows all the secrets of the Universe he tried to understand, in the hands of a Spirit that can be happy that Spirit created Men such as he.

Chapter 6

Yoichiro Nambu

Louis Clavelli

Tufts University and University of Alabama

Of the physicists born in the twentieth century one of the greatest passed away this month, Yoichiro Nambu. He was a legend in the Physics Department at the University of Chicago since his arrival in 1954 in the last days of Enrico Fermi. The department noted in announcing the sad event that Nambu was known not only for his brilliance in physics but also for his deep humanity.

His work in physics is now complete but the inspiration he gave to his academic children and grandchildren lives on. It remains for his former students to provide some details and keep his memory alive. I was there working in his group for six years in the early to mid-sixties. Those were his best and most productive years from the seminal work on spontaneous symmetry breaking to the discovery of the current gauge theory of strong interactions and culminating a few years later in his recognition of dual models as a theory of relativistic strings.

Unfortunately, due perhaps to Nambu's modest demeanor, I and most of his students were slow to realize the full profundity of his ideas. We were there when he told us that he felt the strong interactions were an SU(3) gauge theory acting on a triplet of quarks. The motivation was that this would ensure the ground states were color singlets. This seemed somewhat heuristic to us unexperienced students. Nevertheless, we pored over the requisite books on field theory and particle physics and derived excitement from the weekly "Cyclotron Seminars" which featured the latest theoretical and experimental efforts from around the world. I eventually wrote a dissertation on current algebras and K meson decays following Nambu's papers on the Partial Conservation of the Axial Current (PCAC) and went on to a post-doc position at Yale. A truism about Nambu's ideas spread through the physics community: the importance of his insights could only be appreciated five years later.

In the Chicago physics of the 1960's Nambu served as a bridge between two important camps. The ideology of the conservative camp was that new physics should only be invented when demanded by the data. The main players in this camp were Gregor Wentzel, the co-author of the world famous WKB method, Richard Dalitz, inventor of the Dalitz Plot, and Reinhard Oehme, the local expert in mathematical physics and analyticity. When exposed to some speculative theory in the Cyclotron Seminar, Dalitz once remarked "if he is right we will have to listen to him". The implication was that one should not waste time on the idea until forced to.

The liberal camp was defined by the idea that one should constantly push the envelope exploring what new theories might still be consistent with the data even if not required. This camp was primarily made up of J.J. Sakurai and Peter Freund. Once when Sakurai was presenting the results of some novel assumption, he was asked "Why do you want to assume that?" Sakurai immediately replied "Why not?"

Nambu sympathised with the liberal camp but also saw that speculations were much more interesting if they provided some aesthetic or phenomenological advantage. It was from this point of view that he pursued the "partial conservation of the axial current" (PCAC) and developed the insight with Goto that the dual models of particle interactions were, at a fundamental level, a theory of relativistic strings.

Nambu's style in directing students was to suggest promising areas of research but to encourage maximum self-reliance. He never insisted his students give regular reports on their progress. However, when they felt they had reached a dead end and called for an appointment with him, he would invariably invite them to come immediately up to his office. This worked well for some students but was a disaster for others. One of them languished for ten years before finally getting his doctorate.

Nambu had great human empathy. Once, one of his students was hospitalized after having been mugged on the south side of Chicago. Nambu was there to see him in the hospital and even supported him on his shoulder down to the X-ray facility. One of his most brilliant students rose rapidly in the academic ranks after his PhD but soon succumbed tragically to alcoholism. He returned jobless to Chicago and Nambu on several occasions helped him from his own pocket.

Nambu's sympathy extended to humanitarian causes beyond the physics world. On one occasion a theory student asked him to put his name down in support of a demonstration against the production of napalm to be dropped on Vietnam. Nambu acquiesced but, as a result, he came under FBI investigation and was only saved from deportation by the intervention of senior members of the national physics community. Partially to protect himself from this type of witch hunt, Nambu became a US citizen in 1970.

Nambu's lectures might not have been as polished as those of some of the other professors but they inspired his students to burn the midnight oil working out the necessary details. He would not have been sympathetic to the current academic fashion that holds that a good teacher should be able to pour knowledge into his students with minimal effort on their part.

The depth of Nambu's curiosity extended far beyond natural science. He talked to me once about John Polkinghorne who abandoned a distinguished career in theoretical physics to become a priest in the English church studying ancient greek in order to better analyze what happened 2000 years ago.

He and I last met in 2003 in Gainesville Florida. Nambu no longer casts a shadow on the streets of Kyoto and Chicago but he still casts a long shadow in physics theory.

Chapter 7

Yoicliro Nambu: Remembering an unusual physicist, a mentor and a friend*

G. Jona-Lasinio

Dipartimento di Fisica and INFN,
Università di Roma La Sapienza,
Piazza A. Moro 2, 00185 Roma, Italy

I was lucky to meet Yoichiro Nambu at the beginning of my scientific activity. The experience of working with him influenced my subsequent research and in the following I will try to convey what he transmitted to me. It was also a friendship that continued for decades in spite of the rare occasions to meet after our collaboration.

1. Chicago, spontaneous symmetry breaking, analogies

I arrived in Chicago in September 1959 as a research associate in the theory group of the Fermi Institute. I had an invitation of Herbert Anderson, at that time director of the Institute, who had spent one year in Rome where he had given a course on elementary particle physics. I had written, together with Ugo Amaldi, the notes of this course and this was how I met him. In the summer of 1959 Nambu, on his way to the Kiev Conference, had given a seminar in Rome on his work on Green's functions in quantum field theory. I attended the seminar and in my answer to Anderson I expressed the wish to work with Nambu.

Heisenberg was probably the first to consider spontaneous symmetry breaking (SSB) as a possibly relevant concept in particle physics [1] but he chose the wrong symmetry to be spontaneously broken. It was the intuition of Nambu to catch the analogy between chiral symmetry and gauge invariance in the theory of superconductivity of Bardeen, Cooper and Schrieffer.

The prehistory of this turning point has been recounted by Nambu in the introduction to his collected papers [2]. "*I was very much disturbed*" tells Nambu "*by the fact that their (BCS) wave function did not conserve electron number. It did not make sense, I thought, to discuss electromagnetic properties of superconductors with such an approximation. At the same time I was impressed by their boldness, and tried to understand the problem. I ended up becoming captive of BCS theory.*"

Nambu's paper on superconductivity [3] was a key step as it reformulated BCS theory in a field theoretic language in which the analysis of gauge invariance became particularly transparent and the analogy with chiral invariance could be formulated clearly. Furthermore, simply on the basis of invariance arguments, Nambu realized that zero-mass particles, later called Nambu–Goldstone bosons, were associated to SSB [4].

Nambu and I have reported on previous occasions how the model, now known with the acronym NJL, developed. I will not repeat the story here but refer e.g. to [5] and [6] where I elaborated on cross-fertilization and analogies in theoretical physics.

To understand the impact of working with Yoichiro I must emphasize that my education in physics in Rome had been phenomenologically oriented while my inclinations pushed me towards a more formalized theoretical approach. The type of questions which interested me were of a more foundational character and sometimes I had the feeling that phenomenology was not my natural habitat in spite of the excellent instruction I had received. Yoichiro somehow legitimated my inclinations. I learnt from him two important principles: never be afraid of thinking unconventionally; analogies are a powerful source of ideas.

We worked together intensely for fifteen months but at the end of 1960 I came back to Italy mainly because of family reasons in spite of an offer to remain at the Fermi Institute. The director said to me "*For your work it would be much better to stay here.*" I believe that Yoichiro was behind the proposal, with him we understood each other very quickly and I had developed a kind of elective affinity for his way of thinking. To interrupt the collaboration with Yoichiro was very frustrating and made me very unhappy. Looking retrospectively however this pushed me to acquire a greater independence in my research starting from the new horizons that had opened up in Chicago.

Back home I tried to make good use of what I had learned. In this connection I mention three works in which analogies were the leading thread. With two of my former students we made an analysis of gauge invariance in systems of non-relativistic bosons like superfluid helium, also a case of SSB, inspired by Yoichiro's article on superconductivity, obtaining new interesting sum rules [7]. Then I tried to exploit in a series of papers the formal similarity in structure of quantum field theory and statistical mechanics. In the first of these papers I introduced the so-called effective action, the generating functional of the one-particle irreducible amplitudes, which became popular and a standard tool in the study of spontaneous symmetry breaking [8]. In a work in collaboration with Carlo Di Castro we imported the field theoretic renormalization group into the study of critical phenomena [9], a work especially appreciated by Russian physicists. This was two years before the celebrated papers of Ken Wilson. In those years, the sixties, Yoichiro was permanently in my mind and the question was always what he would think of my work.

2. A long intermission

After Chicago for several decades Yoichiro and I met very seldom but we continued to correspond. Due to my progressive shift towards statistical mechanics we did not attend the same conferences and our relationship was based essentially on a feeling of friendship rather than on scientific exchanges. He visited Rome in 1962 on vacation with his family and his son John told me recently that he has still vivid and pleasant memories of that trip. Then we met again in Italy in 1963 and in Chicago in 1967 that I visited on my way back from a summer at SLAC. At that time he was working on equations for wave functions with infinitely many components while I was still exploiting the analogies between quantum field theory and statistical mechanics.

The next important occasion was his 65th birthday in 1986. Peter Freund and Reinhard Oehme organized a meeting in honor of Yoichiro and asked for contributions for an issue of Progress of Theoretical Physics. I thought that I should contribute something connected with SSB. At that time I was interested in explaining the existence of chiral molecules, the so-called Hund paradox, and had shown in collaboration with Pierre Claverie that an explanation, consistent with experimental data, could be SSB. In fact we never observe isolated molecules so that collective effects are possible. The contribution we submitted [10] showed that the argument could be extended to explain several other molecular phenomena which take place in a semiclassical limit once one takes into account the interaction of a quantum system with the environment. These ideas were later appreciated by Arthur Wightman [11] and the subject is one to which I return intermittently due to its general significance also for quantum measurement theory. It is the long time tail of my collaboration with Yoichiro.

On that occasion with Yoichiro and his wife we recalled the old Chicago days and he made a surprising statement: "*At that time I was very insecure in my work.*" This is an interesting flash on the intellectual itinerary of Yoichiro.

I visited again Chicago ten years later in 1996 during a tour in the US. I remember talking with Oehme about the possibility of a Nobel Prize to Yoichiro and he explained that there was some opposition on the part of condensed matter physicists who considered SSB a common phenomenon in their field so that could not be considered a new discovery. Apparently they were overlooking that in particle physics it had been a major turning point and that only after the transfer into particle physics the generality and key role of this concept was recognized.

I had hoped to see Yoichiro in Rome for my 70th birthday. He was expected to attend a meeting organized by my friends for the occasion but at the last moment he could not come. However he contributed an important paper [12] to an issue of the Journal of Statistical Physics dedicated to this event and sent a warm letter to be read publicly for which I was very grateful.

3. The last years

The Nobel Prize, overdue in my opinion, was finally awarded to Yoichiro in 2008. Due to this event our relationship became again very close and intense as he asked me to represent him in Stockholm. The story went as follows. Obviously I sent him immediately a message of congratulations but for two weeks he did not react. Then he sent me the following short message:

> "*It hurts me that your name appears only in the papers they quoted. I am not going to Stockholm because of my wife's health conditions and mine. But I hope to see more of you.*"

Some days later he sent a new message with the proposal of replacing him for the Nobel lecture but emphasizing that I should be free to say no, he would understand. I greatly appreciated the messages and felt honored by the proposal, my admiration for Yoichiro was unconditional. Besides it was an occasion for me to revisit my past in particle physics. Yoichiro said that he would send a text for the lecture but what he actually sent me was material that he had used in previous lectures and a beginning of the Nobel lecture devoted mainly to his biography. This material was very heterogeneous and not easy to use in a thirty-minute lecture which should illustrate the impact of SSB in the evolution of particle physics. Luckily he said "*use it your own way.*" So I prepared the slides "*my own way,*" proposed a title and submitted them to him. The gratifying answer was "*You have done a better job than I!*"

One year later we met in Kyoto for a meeting in his honor and we had a chance to talk at length. At this meeting I presented some less conventional examples of SSB among which SSB in statistical systems out of equilibrium. In particular I discussed a very simple toy model which showed that SSB could take place in non-equilibrium even if impossible in equilibrium. This model was known to statistical physicists as a model for a traffic jam but also, in a suitable interpretation, as a case of spontaneous CP violation. I suggested that the different behavior of SSB out of equilibrium could be of interest in cosmology in connection with the matter–antimatter asymmetry problem. I sent to Yoichiro a paper that I wrote after the meeting [13] and here is his concise reaction:

> "*...The traffic jam problem is interesting. Recently I actually thought of the traffic problem while driving on Chicago Expressways....*"

At that time he was spending part of the year in Japan where his wife was receiving medical care but commuting with Chicago was very tiring for him. Later he decided to live permanently in Japan near Osaka. His health was worsening, due to kidney problems which were hereditary in his family but had not bothered him previously, and he needed frequent dialysis.

Yoichiro Nambu was an unusual physicist. When he proposed an idea a frequent natural reaction of the audience was "*How did he think of it?*" He had an associative mind with a remarkable mathematical taste and a special feeling for algebraic structures. An outstanding example is provided by the so-called Nambu mechanics formulated in 1973 in a paper by the title *Generalized Hamiltonian Dynamics* [14]. In the last years he has come back to this subject which was entirely his creation with a recent follow up both in mathematics and physics.

Let me give a few details, this work perhaps is not so widely known as his other contributions. Suppose you have a triplet of canonical variables (x_1, x_2, x_3), that is one in addition to the usual

(p, q) and equations of motion of the form

$$\dot{x}_i = \sum_{j,k} \epsilon_{ijk} \frac{\partial H_1}{\partial x_j} \frac{\partial H_2}{\partial x_k} \tag{1}$$

where ϵ_{ijk} is the Levi-Civita tensor, H_1, H_2 are constants of motion. Then for any function $F(x_1, x_2, x_3)$ we have

$$\dot{F} = \sum_{i,j,k} \epsilon_{ijk} \frac{\partial F}{\partial x_i} \frac{\partial H_1}{\partial x_j} \frac{\partial H_2}{\partial x_k}. \tag{2}$$

This structure can be generalized to any number of variables $x_1, \ldots, x_n$ and functions $H_1, \ldots, H_{n-1}$ and this is what Nambu calls a generalized Hamiltonian dynamics. It provides an extension of Poisson algebras associated to Hamiltonian mechanics. For the case of three variables he gives the example of Euler equations for the rigid body where the variables (x_1, x_2, x_3) are identified with the components of the angular momentum, H_1 is the square of the angular momentum and H_2 the kinetic energy. He then discusses quantization which leads to non-associative algebraic structures.

In his last years Yoichiro was interested in the formulation of hydrodynamics within this framework. This implied an infinite-dimensional generalization of his original scheme. He sent me the slides for two seminars he had given in Osaka, the first one "*A particle physicist's view of fluid dynamics*" (subtitle "*An old sake in a new cup*") and the second *"A new look at fluid dynamics."* Shortly before the second seminar in the fall 2013 I had attended a workshop in Cambridge U.K. where I discovered that physicists of the atmosphere, apparently independently, had made the connection of hydrodynamics with Nambu mechanics and had developed effective numerical algorithms based on it. I communicated the news to Yoichiro who was very pleased by this concrete development and quoted their work [15] in the presentation. His comment on this seminar:

> "*...I just gave a talk at an international symposium on physics, Earth and Space sciences. I meant this as the last talk of my life, and I am relaxed now....*"

As I said Nambu's mechanics is having an impact in mathematics, in algebra in particular. I can refer to a recent paper [16] on Nambu's algebras which deals with the finite-dimensional situation of the original 1973 paper. For the infinite-dimensional situation the experts tell me that there are difficulties, initially unexpected, in developing a general theory [17].

I met Yoichiro for the last time in the summer of 2013 in Osaka where I had the honor to inaugurate the newly instituted Nambu Colloquium. On this occasion we talked about various scientific subjects and he was very lively while the Japanese television NHK was taking a film of our discussion. One topic we discussed was the Titius–Bode law of planetary orbits about which he was thinking in his usual original way. Afterwards for a while we continued by mail.

On the last day of my visit I saw him at his home in Toyonaka near Osaka. He was not feeling well and could not come to the University. It was a moving moment for both of us. May be we had a premonition that this was the last time. In a message I received shortly after, he wrote:

> "*...it was a moving moment to see you again. It was fortunate of me to have got to know you. I have to confess that you are the one most close to me in terms of physical thinking, and most comfortable to talk to....*"

This is the friend I lost.

References

[1] W. Heisenberg, W. H. Dürr, H. Mitter, S. Schlieder, K. Yamazaki, *Zeit. f. Naturf.* **14**, 441 (1959).

[2] Y. Nambu, Research in elementary particle theory, in *Broken Symmetries*, eds. T. Eguchi, K. Nishijima, World Scientific, 1995.

[3] Y. Nambu, *Phys. Rev. Lett.* **4**, 380 (1960).

[4] Y. Nambu, *Phys. Rev.* **117**, 648 (1960).

[5] Y. Nambu, Ref. [2]; Nobel Lecture, *Rev. Mod. Phys.* **81**, 1015 (2009).

[6] G. Jona-Lasinio, Cross fertilization in theoretical physics: The case of condensed matter physics and particle physics, in *Highlights in Mathematical Physics*, eds. A. Fokas, J. Halliwell, T. Kibble, B. Zegarlinski, Am. Math. Soc., 2002.

[7] F. De Pasquale, G. Jona-Lasinio, E. Tabet, *Annals of Physics* **33**, 381 (1965).
[8] G. Jona-Lasinio, *Nuovo Cimento* **34**, 1790 (1964).
[9] C. Di Castro, G. Jona-Lasinio, *Phys. Letts.* **29A**, 322 (1969).
[10] G. Jona-Lasinio, P. Claverie, *Prog. Theor. Phys. Suppl.* **86**, 54 (1986).
[11] A. S. Wightman, *Nuovo Cimento B* **110**, 751 (1995).
[12] Y. Nambu, *J. Stat. Phys.* **115**, 7 (2004).
[13] G. Jona-Lasinio, *Progr. Theor. Phys.* **124**, 731 (2010).
[14] Y. Nambu, *Phys. Rev. D* **7**, 2045 (1973).
[15] V. Lucarini, R. Blender, S. Pascale, F. Ragone, J. Wouters, C. Herbert, *Rev. Geophys.* **52**, 809 (2014), arXiv:1311.1190.
[16] N. Cantarini, V. Kac, Classification of linearly compact simple Nambu–Poisson algebras, arXiv:1511.04957.
[17] A. De Sole, private communication.

Chapter 8

Nambu–Goldstone theorem and spin-statistics theorem

Kazuo Fujikawa

RIKEN Nishina Center, Wako 351-0198, Japan

On December 19–21 in 2001, we organized a yearly workshop at Yukawa Institute for Theoretical Physics in Kyoto on the subject of "Fundamental Problems in Field Theory and their Implications". Prof. Yoichiro Nambu attended this workshop and explained a necessary modification of the Nambu–Goldstone theorem when applied to non-relativistic systems. At the same workshop, I talked on a path integral formulation of the spin-statistics theorem. The present essay is on this memorable workshop, where I really enjoyed the discussions with Nambu, together with a short comment on the color freedom of quarks.

1. Workshop at Yukawa institute

In the community of particle theorists in Japan, Yukawa Institute has a special meaning, more so in the old days when I participated in workshops at Yukawa Institute often. The workshop in 2001 was probably the last one I participated at Yukawa Institute. The last one since the decision making committee of Yukawa Institute started to disfavor general titles such as "Fundamental Problems in Field Theory" for a workshop. They preferred more specific titles than a general title, although this principle does not appear to be so strict nowadays. But a more fashionable title such as "String Theory and Field Theory" appears to be favored by the committee.

In any case, this workshop at Yukawa Institute was a memorable one for me, since Prof. Nambu and myself stayed at the same hotel,

Holiday Inn Kyoto, which was located at the north of the campus of Kyoto University. Incidentally, this hotel is no more there since it was closed some time ago. We walked together from the hotel to Yukawa Institute which took about 20 minutes. It is usually very chilly in winter in Kyoto but it was rather mild at that time. I was amazed that Nambu walked quite quickly. Hereafter I just call "Nambu", although I always used "Professor Nambu" in front of him in his life time. My students at University of Tokyo often said that I walked quickly, but Nambu was walking really swiftly at the age of 80. I heard before that Nambu once had some problem with his legs, but this time his health was excellent. On the way from the hotel to Yukawa Institute we discussed various subjects. I recognized that he wanted young Japanese theorists to compete well with counterparts from US and other countries. In this sense he was a "nationalist", but it is often said that a true internationalist is also a good nationalist. Nambu belonged to the old generation of Japanese who respected the elder people. This made me quite nervous when I talked to him since my postdoc days at Chicago; I was always careful about the usage of the Japanese language since the Japanese language contains many honorific words which should be used when you talk to those senior to you. I once heard a story about how Nambu respected and was afraid of Prof. Z. Koba. Koba was senior to Nambu from the days of University of Tokyo just after the Second World War. One day Koba visited Nambu in Chicago from Copenhagen; it is said that Nambu was extremely tense during the visit of Koba by showing Chicago to him, and when Koba left O'Hare Airport to Europe Nambu was really relieved and sighed with relief saying "It's over". In any case, I always presumed that Nambu also expected young Japanese to be similar to him.

But at Kyoto, I felt more relaxed with him partly because he was getting old and thus I enjoyed our conversations on the way to Yukawa Institute from the hotel.

2. Nambu–Goldstone theorem

Nambu mainly played a role of an observer at the workshop but gave a short talk on the Nambu–Goldstone theorem in the presence of a chemical potential [1]. He started his talk by saying that he recently received a letter from Miransky in Ontario, Canada, about a modification of the Nambu–Goldstone theorem for non-relativistic systems [2]. See also [3]. The model Lagrangean that Miransky and Shovkovy and also Nambu discussed is for a complex $SU(2)$ doublet (the K meson system),

$$\Phi(x) = \begin{pmatrix} K^- \\ \bar{K}^0 \end{pmatrix}, \tag{1}$$

and an explicit form of the Lagrangean

$$\begin{aligned} \mathcal{L} = (\partial_0 + i\mu)\Phi(x)^\dagger(\partial_0 - i\mu)\Phi(x) - \partial_k\Phi(x)^\dagger\partial_k\Phi(x) \\ - m^2\Phi(x)^\dagger\Phi(x) - \lambda(\Phi(x)^\dagger\Phi(x))^2 \end{aligned} \tag{2}$$

which is close to the Standard Model Lagrangean with $SU(2) \otimes U(1)$ symmetry. But we now have a chemical potential μ which is introduced to the Lgarangean as a constant time component of a virtual gauge field coupled to the particle number. The chemical potential, which breaks C, CP and CPT, introduces a mass splitting among the particles and antiparticles, $m \to m \pm \mu$. At a certain value of the chemical potential, the lighter ones (for example, the particles) become close to be massless and thus the spontaneous symmetry breaking is triggered. As this qualitative picture implies, only the two massless Nambu–Goldstone modes appear despite that the symmetry breaking spoils 3 generators in $SU(2) \otimes U(1)$ by preserving the electric charge operator formed by a linear combination of a generator of $SU(2)$ and the generator of $U(1)$. The ordinary rule, which states that the same number of massless Nambu–Goldstone bosons as the number of generators contained in G/H appear when the global symmetry group G is spontaneously broken to the symmetry of the ground state H, is not valid any more.

Nambu explained this phenomenon as follows: [1]

> Suppose a broken charge Q develops a vacuum expectation value $\langle Q \rangle = C$. If two other charges Q_i and Q_j are such that their commutator $[Q_i, Q_j] = iQ$, then their corresponding zero modes Z_i and Z_j behave like canonical conjugates of each other: $[Z_i, Z_j] = iC$. Hence they belong to the same dynamical degree of freedom, and the number of Nambu–Goldstone bosons is thereby reduced by

one per each such pair. The dispersion law $\gamma = 2$ (in $\omega \propto |k|^\gamma$) is obtained by a more detailed analysis.

This is a clear statement. During the lunch time at a restaurant outside the campus of Kyoto University after the talk by Nambu, several participants including Nambu and myself continued the discussion. In particular, we discussed much on why the degree of freedom is reduced in a non-relativistic system, although Nambu told us that Nielsen and Chandha [4] gave an analysis in the past. I convinced myself of this reduction on the basis of Bjorken–Johnson–Low (BJL) prescription at the time of discussion: BJL states that the relation for a complex scalar field $\phi(x)$,

$$\int d^4x e^{ikx} \langle T\phi^\dagger(x)\phi(y)\rangle = \frac{1}{\omega^2 - \vec{k}^2} \tag{3}$$

implies $[\phi^\dagger(x), \phi(y)]\delta(x^0 - y^0) = 0$ and $[\partial_0\phi^\dagger(x), \phi(y)]\delta(x^0 - y^0) = i\delta^4(\vec{x} - \vec{y})$ and thus there exist two real freedoms in $\phi(x)$, while

$$\int d^4x e^{ikx} \langle T\phi^\dagger(x)\phi(y)\rangle = \frac{1}{\omega - \vec{k}^2} \tag{4}$$

implies $[\phi^\dagger(x), \phi(y)]\delta(x^0 - y^0) = \delta^4(\vec{x} - \vec{y})$ and thus only one real freedom is contained in $\phi(x)$.

Note that (3) and (4) are expressions in the low energy limit, $\omega \sim 0$ and $|\vec{k}| \sim 0$, and thus the application of BJL limit which is related to short distance behavior is somewhat subtle.

3. Spin-statistics theorem

The spin-statistics theorem states that particles with spin half-odd integers follow the Fermi statistics and those with spin integers follow the Bose statistics. Pauli formulated this theorem on the basis of the following conditions in Lorentz invariant local field theory: [5]

1. Positive energy condition, namely, the energy eigenvalues are bounded from below;
2. Causality in the sense that field variables either commute or anti-commute for the space-like separation;

3. Positive norm condition, namely, no negative probability appears.

I was at that time interested in formulating this theorem in the framework of path integral. The most difficult part in this attempt was how to formulate the positive energy condition. Pauli in his very original formulation used an explicit form of the energy-momentum tensor, but it is not elegant even in the operator formulation. In the operator treatment, Lüders and Zumino [6] later used the spectral condition in the Wightman formalism to improve the original formulation by Pauli. In path integral, the explicit construction of the energy-momentum operator, although not impossible, reduces the treatment to that in operator formalism and the characteristic aspect of the path integral is lost.

I thus proposed to use the Feynman's $i\epsilon$ prescription as a postulate of the positive energy condition [7]. For example, Feynman's prescription means that we should use

$$iS_F(x-y) = \int \frac{d^4p}{(2\pi)^4} \frac{i}{\not{p} - m + i\epsilon} e^{-ip(x-y)} \tag{5}$$

with the replacement

$$m \to m - i\epsilon \tag{6}$$

using an infinitesimal positive ϵ. As is well-known, this prescription specifies that a particle with positive energy propagates forward in time and a particle with negative energy propagates backward in time, which is interpreted as an antiparticle. In either case, the net flow of energy looked at any fixed time slice of space-time is always positive *regardless the assignment of statistics*. This condition is thus used as a positive energy condition. Moreover, the $i\epsilon$ prescription which is often regarded as an *ad hoc* procedure in path integral acquires a more solid physical ground. I believed that this interpretation is physically very nice.

I started my talk on this idea at the workshop by saying that the spin-statistics theorem of Pauli is one of the most fundamental theorems in quantum physics. I was talking while looking at the direction of Nambu, and to my surprise, Nambu expressed an impression of disagreement with my statement. I was embarrassed but I could not understand the reason of his disagreement for many years. But quite recently, by accident I encountered a paper by Nambu and Han written in 1974 [8]; this paper is on a detailed account of the history of color freedom of quarks and its physical meaning after the modern version of color was introduced. The naming of QCD by Gell-Mann was about to be established, and people did not pay enough attention to the original paper by Han and Nambu in 1965 [9] which introduced the Yang–Mills field with $SU(3)$ as the force field among quarks for the first time.

Nambu and Han discussed in some detail the possible physical relevance of the integrally charged quarks they proposed. At the same time they emphasized the equivalence of various past proposals such as $SU(3)$ gauge freedom and the idea of parafermi statistics with order 3 introduced earlier by Greenberg [10], when it comes to the avoidance of the difficulty of the symmetric ground state of 3 quarks in the S-state. Their argument is based on the analysis of Ohnuki and Kamefuchi [11] using the idea of Klein transformation; this analysis implies the non-uniqueness of Fermi and Bose statistics as the mathematically consistent assignments of statistics to elementary particles. I thus thought that I now understand the disagreement of Nambu on my statement on the importance of spin-statistics theorem of Pauli at the workshop.

But I still think that the absence of number systems other than complex numbers for bosons and Grassmann numbers for fermions may imply that the parafermi statistics cannot be formulated in relativistic path integrals.

4. Epilogue

I visited Prof. Nambu at a hospital located near the Osaka Central Station on November 9, 2014. It was raining outside. He was put on oxygen and already rather weak, but still he briefly explained to me the history of the hospital he was staying at, which is apparently related to the Medical School of Kyoto University. He also told me that Mrs. Nambu was staying at another hospital. But otherwise we were unable to talk about physics, since he told me that it is hard to keep his body straight on the bed. When I told him that I was once hospitalized 3 years ago, he showed some interest and asked me what was the cause of hospitalization. I told him that it was related to the irregularity of the self-immune system of the body. Otherwise, not much conversations. I left his hospital room after a short stay, since he repeated that he felt tired.

I was sad since I remembered a very lively discussion on what he was working at the occasion of his public lecture at Kyoto University and a small nice workshop at Yukawa Institute on the following day on October 26–27, 2009, after he received the Nobel Prize in 2008. At that time he told me that he was working on the membrane and quizzed me if I can imagine for what purpose it is used. I replied to him that I cannot imagine, then he smiled and said "fluid dynamics!". Nambu had a strong interest in fluid dynamics just as great physicists like Heisenberg. In any case, Nambu tended to ask this sort of question to younger people and enjoyed the response. I felt a bit embarrassed but I told him that the membrane theory when quantized contains area-preserving diffeo which may be related to the incompressibility of the fluid. He also asked me if I bothered to come to Kyoto from Tokyo to attend his lecture and the small workshop and expressed his gratitude. He was a real gentleman.

I really wish that he kept his health for several more years and I could have more occasions to discuss with him on various aspects of physics.

References

[1] Y. Nambu, *Symmetry Breaking by a Chemical Potential*, Soryushiron Kenkyu, Vol. **105**, No. 4, Page D24–D25 (2002). Available at the web site, CiNii Articles-Soryushiron Kenkyu.
[2] V.A. Miransky and I.A. Shovkovy, *Phys. Rev. Lett.* **88**, 111601 (2002).
[3] T. Schaefer, D.T. Son, M.A. Stephanov, D. Toublan, and J.J.M. Verbaarschot, *Phys. Lett. B* **522**, 67 (2001).
[4] H.B. Nielsen and S. Chandha, *Nucl. Phys. B* **105**, 445 (1976).
[5] W. Pauli, *Phys. Rev.* **58**, 716 (1940).
[6] G. Lüders and B. Zumino, *Phys. Rev.* **110**, 1450 (1958).

[7] K. Fujikawa, *Int. J. Mod. Phys. A* **16**, 425 (2001).
[8] Y. Nambu and M.-Y. Han, *Phys. Rev. D* **10**, 674 (1974).
[9] M.-Y. Han and Y. Nambu, *Phys. Rev.* **139**, B1006 (1965).
[10] O.W. Greenberg, *Phys. Rev. Lett.* **13**, 598 (1964).
[11] Y. Ohnuki and S. Kamefuchi, *Prog. Theor. Phys.* **50**, 258 (1973).

Chapter 9

Pre-string theory

Paul H. Frampton

University of North Carolina
paul.h.frampton@gmail.com

In this note, I recollect a two-week period in September 1968 when I factorized the Veneziano model using string variables in Chicago. Professor Yoichiro Nambu went on to calculate the N-particle dual resonance model and then to factorize it on an exponential degeneracy of states. That was in 1968 and the following year 1969 he discovered the string action. I also include some other reminiscences of Nambu who passed away on July 5, 2015.

Here my purpose is to describe early work on string theory immediately following the appearance of the Veneziano model. In unpublished work in 1968, a multiparticle generalization was factorized as a string theory, and, in further unpublished work in 1969, the bosonic string action was discovered at the University of Chicago by Professor Yoichiro Nambu who passed away on July 5, 2015. As his postdoc, I contributed to the very beginning of string theory in answering a question by Nambu and factorizing the basic Veneziano model during September 9–23, 1968. After that I was a spectator.

I first became aware of the Veneziano model on Friday, September 6, 1968. I can be certain of the exact date because it was after the ICHEP68 conference in Vienna Aug 28–Sept 5 and before I flew to Chicago for a postdoc on Sept 7. I did not attend the Vienna conference but my research supervisor J.C. Taylor did. As soon as he returned to Oxford, Taylor showed me the Veneziano model which had been widely discussed in Vienna and was a big surprise. For a two-particle scattering with momenta $p_1 + p_2 \rightarrow (-p_3) + (-p_4)$ the proposed model for the scattering amplitude $A(s,t)$ with $s = (p_1 + p_2)^2$ and $t = (p_2 + p_3)^2$ was

$$A(s,t) = \int_0^1 x^{-\alpha(s)-1}(1-x)^{-\alpha(t)-1}dx \tag{1}$$

with $\alpha(x) = \alpha(0) + \alpha' x$. It could be readily checked that this can be written as an infinite tower of resonances in either direct (s) or crossed (t) channels. Remarkably Eq. (1) satisfies all FESRs in the average sense of DHS duality. It is manifestly $s - t$ crossing symmetric and Regge pole behaved in both channels. Such a simple closed solution of all of the FESRs was quite unexpected.

On Monday, September 9, 1968 I reported to start work as a postdoc in the University of Chicago and there I met for the first time Professor Yoichiro Nambu. He was very friendly and welcoming. He was a small man and rather quiet. At that time his English was imperfect but readily comprehensible if one listened carefully. He laughed and smiled a lot.

When he asked what I was interested in, I mentioned the Veneziano model, wrote Eq. (1) on the blackboard, and discussed its properties. Nambu then posed to me a very interesting question about Eq. (1): can the t-dependence be factorized as follows:

$$(1-x)^{-2\alpha' p_2 . p_3} = F(p_2)G(p_3)? \tag{2}$$

This was a question which I could not immediately answer and neither could he. I promised to try and that was my first assignment. He was my first boss in my first job so I was eager to impress him. I spent two weeks day and night wrestling with the impossible-seeming Eq. (2) and returned two weeks later on Monday, September 23 with an explicit solution.

The first step was to write

$$(1-x) = \exp[\ln(1-x)] \tag{3}$$

and to expand the logarithm

$$\ln(1-x) = -\sum_{1}^{\infty} \frac{x^n}{n}. \tag{4}$$

I was familiar with the Baker–Hausdorff theorem which states

$$e^A e^B = e^B e^A e^{[A,B]} \tag{5}$$

providing that the commutator $[A, B]$ commutes with A and B. I was familiar also with the method of solving the quantum harmonic oscillator using operators which satisfy $[a, a^\dagger] = 1$ and a ground state $|0\rangle$ with $a|0\rangle = 0$. To complete the solution therefore I needed an infinite number of oscillators with what would now be called string variables

$$[a_\mu^{(m)}, a_\nu^{(n)\dagger}] = -g_{\mu\nu}\delta_{mn} \tag{6}$$

with ground state $|0\rangle$ satisfying $a_\mu^{(m)}|0\rangle = 0$.

With this background and defining $F(p)$ and $G(p)$

$$F(p) = \exp\left(i\sqrt{2\alpha'}p_\mu \sum_{1}^{\infty} \frac{a_\mu^{(n)} x^n}{\sqrt{n}}\right) \tag{7}$$

$$G(p) = \exp\left(i\sqrt{2\alpha'}p_\mu \sum_{1}^{\infty} \frac{a_\mu^{(n)\dagger}}{\sqrt{n}}\right) \tag{8}$$

it is easily checked that the explicit solution of Eq. (2) is

$$(1-x)^{-2\alpha' p_2 . p_3} = \langle 0|F(p_2)G(p_3)|0\rangle. \tag{9}$$

Nambu seemed pleased and continued to work on the Veneziano model. First he calculated the generalization from 4 to N particles and I helped him check that his results agreed with the explicit proposals in the literature by Bardakci and Ruegg, and by Chan and Tsou. Nambu also calculated the degeneracy $d(N)$ of the level $\alpha(s) = N$ as the number of partitions of N into integers, a problem solved in 1917 by Hardy and Ramanujan, which for large N goes as $d(N) \sim \exp(c\sqrt{N})$.

All of this was well in hand before the end of the calendar year 1968 and anybody else would have published a paper in, say, December 1968. But Nambu had somehow heard rumours that two other groups, one at MIT (Fubini and Veneziano) and another at Berkeley (Bardakci and Mandelstam), had independently discovered the exponential degeneracy.

Nambu seemed to receive adequate satisfaction just from the creativity involved without needing to publish. An earlier example was in 1948 in Tokyo when he calculated the one-loop correction to the electron magnetic moment but did not publish when he read in Time magazine that Schwinger had also done it.

We waited for what seemed forever, actually until April 1969, until two papers, one from MIT the other from Berkeley, arrived. Both had used brute force methods such as multinomial expansions, and had not found Nambu's far simpler string operators. I clearly remember Nambu's reaction was, sitting in his office holding these two long papers with a smile, to say: "Why did they make it so complicated?"

The 1968 results of Nambu remained unpublished. In June 1969 papers by Nielsen and by Susskind appeared saying similar things, not knowing about Nambu's results. Priority for string theory based on publications is now universally attributed to Nambu, Nielsen and Susskind. Nambu did give a talk in June 1969 at a conference in Detroit but if, as he easily could have, he had published a refereed paper six months earlier string theory would have been solely his property.

In July 1969, Nambu came to my office to tell me that the action of the string is the area of its world-sheet. For some reason, I was so skeptical that he bowed, apologized and went away saying he must have made a mistake. But when I learned that the action for a particle is the length of the world-line I realized that it was another brilliant discovery. Just as for the string theory, however, also in the case of the string action Nambu did not write it up until a year later for lectures at the Copenhagen Summer School in 1970. In May 1971, a Japanese physicist named Goto published it and it is generally called the Nambu–Goto action. If he had published a refereed paper about it in 1969, it would be the Nambu action.

After leaving Chicago[a] for CERN in 1970, we stayed in touch. For example in December 2002, he very kindly took me by bus and train from Nagoya to the nearby Meiji Mura architectural park. There was a reconstruction of the Frank Lloyd Wright Imperial Hotel in Tokyo where Nambu said he had stayed as a small boy before the earthquake of 1923, so he must have been less than three years old. There was also a reconstruction of the home of the Japanese writer Natsume Soseki whose autobiographical novel *Grass on the Wayside* Nambu greatly admired.

The last time I met him was at Osaka in March 2010 when Hosotani invited me for a seminar and Nambu came in for the day. He had a permanent office there but came in only very rarely. By coincidence, a number of students were graduating and were delighted to have their picture taken with him and their graduation certificates signed.

The two of us spent a very pleasant couple of hours leafing gradually through a book of his collected publications about which he told many interesting anecdotes. But those publications might represent only the tip of an iceberg. He always worked hard and wrote everything out in his neat handwriting. In his Chicago office in 1968 he already had a lot of notebooks going back decades all lined up on his shelves. Now there must be many more and I wonder what remains undiscovered there.

[a]During the two-year period, we did successfully publish one paper together in a festschrift for Professor Wentzel. The reference is: Y. Nambu and P. Frampton, Asymptotic behavior of partial widths in the Veneziano Model, in *QUANTA, A Collection of Scientific Essays Dedicated to Gregor Wentzel*, University of Chicago Press (1970).

Chapter 10

Some reminiscences from a long friendship

Lars Brink

Department of Fundamental Physics,
Chalmers University of Technology,
S-412 96 Göteborg, Sweden
lars.brink@chalmers.se

1. My first encounters with Yoichiro Nambu

When I was in my first year of graduate studies, our department organised the first Nobel Symposium in physics in the spring of 1968. Essentially all the leading physicists in particle physics were there, and for a young student it was a great time to see and listen to all those famous scientists. The meeting became later very famous for Abdus Salam's contribution [1] that led to his Nobel Prize.

One of the speakers was Yoichiro Nambu. He talked about infinite-component field theories [2], a subject that is now solidly forgotten. I was in a small room in the back recording the sessions but could follow the talks, even though I did not understand much. I could see, however, the respect that the other speakers showed to him.

In the fall of 1969 I visited my advisor who was on sabbatical in Austin for a few months. Nambu came there to give a talk and he then lectured about his factorisation of the Veneziano Model amplitudes. This was certainly a subject I was beginning to be drawn to.

In the summer of 1970 I went to a summer school in Copenhagen mainly because Nambu was going to talk about dual models as it was announced. This was then the subject I had started to work on not telling my advisor. Nambu did not make it there. Much later he told me that his car had broken down on his way home to Chicago to go to Copenhagen. However, his preprint [3] for the meeting had arrived and it was distributed among the participants. This is the preprint in which he suggested the action for the bosonic string which came to be called the Nambu–Gotō action. In the preprint he dared to be braver than in an ordinary paper and there were lots of interesting ideas. He was even comparing the string to the DNA at the end of the preprint. It became eventually a rarity and I used to keep it in a box in the attic of our department until someone cleaned out all my boxes of old preprints. The Nambu paper was the only one I really missed.

2. Zero-point fluctuations of strings

The academic years 1971–73 I spent at CERN as a fellow. I was gradually more and more drawn into the field of 'Dual Models'. In the final year I worked with David Olive on a huge program to compute the correct one-loop graphs which we did by using Feynman's tree theorem. This amounts to sewing tree diagrams together and inserting a projection operator onto the physical states in one of the propagators. Feynman had mentioned this technique at a question session at a conference in Poland in 1963 [4] and we reconstructed it [5] showing that it indeed gives the correct perturbative unitarity and showed that there were no problems with conflicting $i\epsilon$-prescriptions etc.

Having constructed the physical state projection operators in all sectors of the open and closed strings we could not only construct the one-loop diagrams [6] but could also check the gauge properties of the open–closed string (reggeon–pomeron) vertex [7] as well as the fermion emission vertex [8], showing that the Ramond and the Neveu–Schwarz models coupled unitarily to each other, and we wrote a long series of papers that year.

In the summer of 1974 John Schwarz organised a workshop in Aspen on Dual Models and at the same time Murray Gell-Mann organised a workshop on gauge field theories. A number of very famous people were around, and one day Bunji Sakita came up to me and said that Prof Nambu wants to meet with you. I went in his office and introduced myself. I thought he wanted to discuss about my work with David Olive but no. He wanted to congratulate me on my paper [9] with Holger Bech Nielsen on zero-point fluctuations in string theory. Nambu had himself [10] been the first to see that the states in dual models resembled an infinite set of harmonic oscillators. In the bosonic model the mass spectrum was given by the formula

$$\alpha' m^2 = \sum_{n=1}^{\infty} n {a^\dagger}_n \cdot a_n. \tag{1}$$

If we do it in the light-cone gauge, only the transverse oscillators remain and you sum over $d-2$ components. Every oscillator will contribute a zero-point fluctuation of $(1/2)\hbar\omega$. Formally the mass of the lowest lying state would then be

$$\alpha' m_0^2 = \frac{d-2}{2} \sum_{n=1}^{\infty} n. \tag{2}$$

This is clearly a divergent result and must be regularised. We did it by regularising and renormalising the velocity of light which is the speed of the phonons along the string. A simpler way to do it is to use ζ-function regularisation. (This was later pointed out to us by Nando Gliozzi.)

$$\alpha' m_0^2 = \frac{d-2}{2} \sum_{n=1}^{\infty} n^{\zeta}|_{\zeta=1} = -\frac{d-2}{24}. \tag{3}$$

By knowing that the first oscillation is a massless vector particle we could draw the conclusion that $d = 26$.

The fact that the sum of all integers can be regularised to become $-\frac{1}{12}$ was a well-known result in mathematics perhaps going all the way back to Jacobi (and his assistant J. Scherk).

This was a paper that took us a few days to write, but which we were quite proud of and later had a lot of use for. Anyhow this was the paper that Nambu congratulated me for. I think it shows his good taste in physics if I may say so.

3. The σ-model action or how to supersymmetrise the Nambu–Gotō action

In his Copenhagen lecture [3] Nambu described how the spectrum of the states in the Veneziano model could be described by a free bosonic string with a subsidiary condition. Consider the action for a free string

$$I = \frac{1}{4\pi}\int d^2\xi \left(\frac{\partial x_\mu}{\partial \tau}\frac{\partial x^\mu}{\partial \tau} - \frac{\partial x_\mu}{\partial \sigma}\frac{\partial x^\mu}{\partial \sigma}\right), \tag{4}$$

where $\xi^0 = \tau$ and $\xi^1 = \sigma$. This action is conformally invariant. By demanding that the Noether current generating these conformal transformations be zero we can generate the Virasoro conditions which eventually was proven to lead to a unitary spectrum.

The action (4) is not a geometric action so Nambu says that it leads us to construct one out of curiosity. A natural candidate is then the surface area, the two-dimensional world-sheet. The sheet is embedded in the Minkowski space as $x^\mu(\tau,\sigma)$ so the surface element is

$$\begin{aligned} d\sigma^{\mu\nu} &= G^{\mu\nu} d^2\xi, \\ G^{\mu\nu} &= \partial(x^\mu, x^\nu)/\partial(\tau,\sigma), \end{aligned} \tag{5}$$

whereas the line element is

$$\begin{aligned} ds^2 &= g_{\alpha\beta} d\xi^\alpha d\xi^\beta, \\ g_{\alpha\beta} &= \left(\frac{\partial x^\mu}{\partial \xi^\alpha}\frac{\partial x_\mu}{\partial \xi^\beta}\right). \end{aligned} \tag{6}$$

The geometric action is then

$$I = \int |d\sigma^{\mu\nu}\, d\sigma_{\mu\nu}|^{1/2} = \int d^2\xi \sqrt{(\dot{x}x')^2 - (\dot{x}^2 x'^2)}. \tag{7}$$

Since this is reparametrisation invariant on the world-sheet we can impose two conditions such that the equations are linearised and agree with the equations that follow from (4). The subsidiary conditions chosen for this purpose are

$$\frac{\partial x}{\partial \tau} \cdot \frac{\partial x}{\partial \sigma} = 0, \quad \left(\frac{\partial x}{\partial \tau}\right)^2 + \left(\frac{\partial x}{\partial \sigma}\right)^2 = 0, \tag{8}$$

which just amounts to the Virasoro conditions.

In 1974 Paolo Di Vecchio had arrived at Nordita in Copenhagen as an assistant professor and we started to collaborate. One problem we soon started to discuss was how to supersymmetrise the Nambu–Gotō action. At this time supersymmetric field theories were a hot subject and we wanted to use the knowledge from the four-dimensional theories for the supersymmetric string as we called it. Already in 1971 some six months after Pierre Ramond's ground-breaking paper [11] Gervais and Sakita [12] had constructed a free supersymmetric two-dimensional action. We soon realised that by demanding that the Noether current from the superconformal symmetry be zero, we could generate the super-Virasoro conditions. But why should they be zero? In this process we discovered with altogether a full football team of Italians that we could generalise this procedure to infinitely many super-Virasoro algebras [13] out of which two could be represented as extensions of the original ones. In this way we found two new supersymmetric string models [14, 15] but none of them turned out to be close to the one we searched for, a four-dimensional model with a hadronic spectrum. We were still looking for a model for the strong interactions.

Could one add in fermionic degrees of freedom to the Nambu–Gotō action and get the supersymmetric string? One natural attempt was to introduce a supercoordinate instead of the usual coordinate x^μ. That worked at the linearised level. However, we soon found out that to have Grassmann numbers under the square-root sign is not a good idea. Every attempt failed and we were stuck on the problem.

In the beginning of 1976 the first papers on supergravity appeared [16, 17]. In principle we should have understood immediately how to find a reparametrisation invariant string action with Grassmann coordinates, but we were slow to understand it. The formalism with vierbein fields was unknown to us and we were doing other things. Finally in the summer Paolo Di Vecchia, Paul Howe and I got together at CERN for a few days and constructed the corresponding action for a supersymmetric particle, an action that leads to the Dirac equation. We wrote it up with Stanley Deser and Bruno Zumino [18]. I was about to move to Caltech for a year but we found some time in September and during some very hectic days we finally realised how to generalise the Nambu–Gotō action [19]. By introducing a two-dimensional gravity field we could simply rewrite (7) as

$$I = -\frac{1}{2}\int \sqrt{-g}\; g^{\alpha\beta}\, \partial_\alpha x_\mu \partial_\beta x^\mu, \tag{9}$$

where g is the determinant of $g_{\alpha\beta}$.

The extension to a supersymmetric string is then straightforward. One has to extend the two-dimensional reparametrisations to a local supersymmetry and to introduce zweibein fields $e_\mu{}^\alpha$ and their superpartners $\chi_\mu{}^a$ to couple the fermion coordinates $\lambda_\mu{}^a$ in an invariant way. The final action for the superstring is then

$$\begin{aligned} I = \int e \Big[& -\frac{1}{2} g^{\alpha\beta}\, \partial_\alpha x_\mu \partial_\beta x^\mu - \frac{i}{2}\bar{\lambda}_\mu \gamma^\alpha \partial_\alpha \lambda^\mu \\ & + \frac{i}{2}\bar{\chi}_\alpha \gamma^\beta \gamma^\alpha \lambda_\mu \partial_\beta x^\mu + \frac{1}{8}(\bar{\chi}_\alpha \gamma^\beta \gamma^\alpha \lambda_\mu)(\bar{\chi}_\beta \lambda_\mu) \Big], \end{aligned} \tag{10}$$

where e is the determinant of the zweibein.

By choosing gauge fixings properly one can linearise the equations of motion and obtain the super-Virasoro constraints. Finally we got the action we had been looking for such a long time. It was, of course, known since long that the critical dimension for this theory is ten. The easiest way to find it is to consider the mass-square operator which follows from the constraints and commute the zero-point fluctuations.

4. Nambu and the Nobel Prize

In 2001 I got first elected as an adjoint member of the Nobel Committee for physics. The work in the committee is, of course, highly classified but anyone can conclude that Nambu became one of the people that I had to consider. I was fortunate that the year before I got a copy from Valentine Telegdi of the book "Broken Symmetry: Selected Papers of Y. Nambu" edited by T. Eguchi and K. Nishijima [20]. The book contains all the important papers of Nambu up to the 1970's and it also contains some of the talks he had given over the years. As I will indicate later some of these talks were decisive for me to understand his work. The work in the committee is quite lonely. You cannot show an interest in a person outside a small circle. Over the years I had to restrain myself not to ask about Nambu's health too often.

Nambu told me once that he was very mystified when Bob Schrieffer came to give a seminar in Chicago in 1956 just before they were about to publish the famous BCS paper [21]. What happened to gauge invariance? The condensation of the Cooper-pairs clearly leads to a new vacuum that does not conserve the charge. The BCS theory is essentially QED coupled to phonons and if the gauge invariance is broken, how can one trust any calculations? It took Nambu 2.5 years to understand and solve the problem to his satisfaction [22]. In the meantime superconductivity was one of the hottest topics in physics, and since it has a field theoretic background both condensed matter and particle physicists worked on it. Anderson [23] introduced the collective excitations and Bogoliobov [24] introduced his quasi-states but the results were attacked for example by Shafroth who firmly stated that the results cannot be trusted since the calculations are not gauge invariant.

However, the results in the BCS paper were clearly correct. Nambu solved the problem in a tour de force calculation which is one of the toughest to follow in detail. He started by introducing a vacuum with a vacuum expectation value for the electron field $\langle\psi(x)\psi(y)\rangle$ in addition to $\langle\bar{\psi}(x)\psi(y)\rangle$ and then he used a Feynman–Dyson formulation to compute the higher quantum correction. In the end he found that gauge invariance is conserved although in a non-linear fashion. The collective excitations come out as solutions to his vertex equations and all calculations done before can now be shown to be trusted. Unfortunately Shafroth was no longer around to see the solution. He and his wife had crashed with a small postal plane in Australia a month before Nambu's paper was available.

The paper is certainly one of the most important papers of the last century. Not only did it explain superconductivity in a consistent way. It also laid the ground for a lot of the particle physics that would come after it. Nambu was not a scientist that wrote everything on the reader's nose. One has to dig in the paper. One might ask how much of the BEH–effect [25–28] is there? Anderson has often praised Nambu and the superconductivity paper, but alway insisted that Nambu did not have the effect in the paper. It is true that the model set up is non-relativistic since it contains a Fermi surface, but it has particles and antiparticles and gauge invariance. At the end of the paper when Nambu discusses the Meissner effect he writes: "We see that the previous collective state (which corresponds to mass zero, my comment) has shifted its energy to the plasma energy as a result of the Coulomb interaction." This looks to me as very similar to Anderson's ideas some years later [29]. What is clear though is that Nambu had already got his next great idea, namely to use this technique on pion physics. I am sure that somehow he knew that gauge fields could have short range and still be gauge covariant but he did not push it. As the gentleman he was, he never claimed any priority for it either.

After Feynman's and Gell-Mann's introduction of the Conserved Vector Current hypothesis (CVC) [30], the question was if there was a similar relation for the axial vector current. Without a field theory for the strong interaction it was hard to envisage how such a current would look like. However, it was a tempting idea. Within four days in February 1960 Gell-Mann and Lévy [31] and Nambu

[32] published two classical papers that changed the world. Gell-Mann and Lévy set up several models that had Partially Conserved Axial Vector Current (PCAC). They also introduced the non-linear σ-model and what came to be called the Cabibbo angle and computed it correctly. Nambu's paper was very short and to the point. He simply suggested a form for the current (in his notation and normalisation)

$$g_A {\Gamma_\mu}^A(p', p) = g_V \left[i\gamma_5\gamma_\mu F_1(q^2) - \frac{2M\gamma_5 q_\mu}{q^2 + {m_\pi}^2} F_2(q^2) \right] \tag{11}$$

$$F_1(0) = g_A/g_V = F_2(0)$$

$$F_1(q^2) \sim F_2(q^2) \quad \text{for } q^2 \gg {m_\pi}^2,$$

where M is the nucleon mass. The form factors $F_{1,2}$ carry the part of the interaction that is unknown. When the pion mass is zero the current is conserved. In the real life it is partially conserved. The q_μ in the second term shows that the nucleon–nucleon–pion interaction has a derivative coupling, very different from the Yukawa coupling. It also implies that the neutron emits a massless pion which propagates and then decays to $e + \nu$ in the neutron β decay.

From this formula Nambu derives the "Goldberger–Treiman relation" (again in his notations)

$$2Mg_A = 2Mg_V F_2(-{m_\pi}^2) = \sqrt{2} G_\pi g_\pi, \tag{12}$$

where g_π is the pion decay constant and G_π is the pseudoscalar coupling constant. This relation which fit surprisingly well the data had been derived with some violent approximations by Goldberger and Treiman [33] and here it just came out of Nambu's assumption about PCAC.

But where did he get all this from? What more did he have in his hat? For someone reading the paper 50 years later it is incomprehensible. Fortunately he also gave talks about it and at a meeting in Purdue [34] a little later he really expounded his ideas. This was very similar to what he did in his Copenhagen lectures in the sense that he was quite explicit. He once told me that in lecture notes he could be more detailed and also propose wilder ideas. Like at Copenhagen he did not deliver the talk at Purdue. It was delivered by Gianni Jona-Lasinio. These lecture notes somehow disappeared, however, and it was impossible to read them until the book with the selected articles came out. For this volume it was retypeset and it is now a gold mine to Nambu's ideas at that time.

From the start he shows how influenced he was by his understanding of superconductivity. He sets up a table that explains the correspondences between superconductivity and pion physics.

Superconductivity	**Pion physics**
Free electrons	bare fermions (zero or small mass)
Phonon interaction	some unknown interaction
Energy gap	observed mass of nucleon
Collective excitation	meson, bound nucleon pair
Charge	chirality
Gauge invariance	γ_5-invariance (rigorous or approximate)

Now it becomes clear how he thought. He thought of a vacuum with a condensation of fermion–antifermion pairs and with mesons built up by those pairs where the mass of the fermions are a few MeV. This is four years before the quarks but four years after Sakata's work [35] that was influential at the time. However he had realized that the constituent particles should have a small mass. This is much more like the up- and down-quarks developed four years later. He then shows without knowing the details of the strong interactions how the spontaneously broken symmetry is restored in the limit of vanishing pion mass. This is again a non-linear realisation of the symmetry. The new thing is that the spontaneous breaking of the symmetry forces a massless particle into the theory, in this case the pion, what was to be called the Nambu–Goldstone particle. This is different from the spontaneous breaking of the gauge invariance in the superconductivity case, a case I do not find that he comments on. In a stroke of genius he had explained how nuclear physics could work where the energy scales are of the order of GeV. With a spontaneous breaking the chiral symmetry and with a small explicit breaking of this with a small quark-mass, one can get an enough long range strong force that binds the nuclei together.

After these ground-breaking papers Nambu wrote a series of papers with Jona-Lasinio [36, 37] and others [38] to extract useful information about the pion physics even though the exact structure of the force was not known. This was very important at the time and became even more important when QCD came around.

Finally in 2008 the Nobel Committee was ready to award the Nobel Prize to Nambu. The work started early in the year and I was nervous about his health. He was 87 years old. I met Bob Wald at a meeting in January and I asked him thoroughly about his colleagues. Even though I was interested in their health it was mainly Nambu's health I was interested in. Finally I asked about that and got a positive answer but there was still some nine months to go.

On October 7 we had the final vote in the Academy of Sciences and after that we went to our secret telephone and called Nambu. It was very early in Chicago but finally he answered. I am sure he was at sleep not worrying about telephone calls from Sweden. First the permanent secretary of the Academy talked to him and told him the good news and then I got the receiver. Then I lost my voice and I got so moved so I could only say "This is Lars. Congratulations!"

I had a long-standing dream that I should be giving the speech at the ceremony and when I turned to Nambu I should be able to address him in Japanese. In the end he could not make it to the ceremony in Stockholm since his wife was not well enough. However, there were two more Japanese laureates, Makoto Kobayashi and Toshihide Maskawa, so I did prepare the last part of the talk, the address to the laureates in Japanese. I had very good help by a local professor of Japanese. I did mention this to Murray Gell-Mann on the phone who immediately gave me a long course in Japanese pronunciation, which did not make me less nervous. In the end it went quite well and the only problem arose when I finished, since only the laureates and I knew that the speech was over. After a long minute the king raised and the ceremony could go on.

Since there was a time gap of seven hours between Stockholm and Chicago, the video of my talk, that was given in Swedish except the last part, could be transmitted and subtitles could be introduced, so at the ceremony in Chicago the audience could even understand the Swedish part of the talk.

One dream I had had for so long had finally come true.

References

[1] A. Salam, Weak and electromagnetic interactions, *Conf. Proc. C* **680519**, 367 (1968).
[2] Y. Nambu, Relativistic groups and infinite-component fields. in N. Svartholm, *Elementary Particle Theory. Proceedings of The Nobel Symposium* Lerum, Sweden, Stockholm, 1968, 105–117.
[3] Y. Nambu, Duality and Hadrodynamics, Notes prepared for the *Copenhagen High Energy Symposium*, Aug. 1970.
[4] R. P. Feynman, Quantum theory of gravitation, *Acta Phys. Polon.* **24**, 697 (1963).
[5] L. Brink and D. I. Olive, The physical state projection operator in dual resonance models for the critical dimension of space-time, *Nucl. Phys. B* **56**, 253 (1973).
[6] L. Brink and D. I. Olive, Recalculation of the unitary single planar dual loop in the critical dimension of space time, *Nucl. Phys. B* **58**, 237 (1973).
[7] L. Brink, D. I. Olive and J. Scherk, The gauge properties of the dual model pomeron-reggeon vertex — their derivation and their consequences, *Nucl. Phys. B* **61**, 173 (1973).
[8] L. Brink, D. I. Olive, C. Rebbi and J. Scherk, The missing gauge conditions for the dual fermion emission vertex and their consequences, *Phys. Lett. B* **45**, 379 (1973).
[9] L. Brink and H. B. Nielsen, A simple physical interpretation of the critical dimension of space-time in dual models, *Phys. Lett. B* **45**, 332 (1973).
[10] Y. Nambu, Quark model and the factorization of the Veneziano amplitude, in *Proceedings, Conference On Symmetries* Detroit 1969.
[11] P. Ramond, Dual theory for free fermions, *Phys. Rev. D* **3**, 2415 (1971).
[12] J. L. Gervais and B. Sakita, Field theory interpretation of supergauges in dual models, *Nucl. Phys. B* **34**, 632 (1971).

[13] M. Ademollo *et al.*, Supersymmetric strings and color confinement, *Phys. Lett. B* **62**, 105 (1976).
[14] M. Ademollo *et al.*, Dual string with U(1) color symmetry, *Nucl. Phys. B* **111**, 77 (1976).
[15] M. Ademollo *et al.*, Dual string models with nonabelian color and flavor symmetries, *Nucl. Phys. B* **114**, 297 (1976).
[16] D. Z. Freedman, P. van Nieuwenhuizen and S. Ferrara, Progress toward a theory of supergravity, *Phys. Rev. D* **13**, 3214 (1976).
[17] S. Deser and B. Zumino, Consistent supergravity, *Phys. Lett. B* **62**, 335 (1976).
[18] L. Brink, S. Deser, B. Zumino, P. Di Vecchia and P. S. Howe, Local supersymmetry for spinning particles, *Phys. Lett. B* **64**, 435 (1976).
[19] L. Brink, P. Di Vecchia and P. S. Howe, A locally supersymmetric and reparametrization invariant action for the spinning string, *Phys. Lett. B* **65**, 471 (1976).
[20] T. Eguchi and K. Nishijima, *Broken Symmetry: Selected Papers of Y. Nambu*, World Scientific, 1995.
[21] J. Bardeen, L. N. Cooper and J. R. Schrieffer, Theory of superconductivity, *Phys. Rev.* **108**, 1175 (1957).
[22] Y. Nambu, Quasiparticles and gauge invariance in the theory of superconductivity, *Phys. Rev.* **117**, 648 (1960).
[23] P. W. Anderson, Coherent excited states in the theory of superconductivity: Gauge invariance and the Meissner effect, *Phys. Rev.* **110**, 827 (1958).
[24] N. N. Bogoliubov, On a new method in the theory of superconductivity, *J. Exptl. Theoret. Phys.* **34** (1), (1958).
[25] F. Englert and R. Brout, Broken symmetry and the mass of gauge vector mesons, *Phys. Rev. Lett.* **13**, 321 (1964).
[26] P. W. Higgs, Broken symmetries and the masses of gauge bosons, *Phys. Rev. Lett.* **13**, 508 (1964).
[27] A. A. Migdal and A. M. Polyakov, Spontaneous breakdown of strong interaction symmetry and the absence of massless particles, *Sov. Phys. JETP* **24**, 91 (1967) [*Zh. Eksp. Teor. Fiz.* **51**, 135 (1966)].
[28] G. S. Guralnik, C. R. Hagen and T. W. B. Kibble, Global conservation laws and massless particles, *Phys. Rev. Lett.* **13**, 585 (1964).
[29] P. W. Anderson, Plasmons, gauge invariance, and mass, *Phys. Rev.* **130**, 439 (1963).
[30] R. P. Feynman and M. Gell-Mann, Theory of Fermi interaction, *Phys. Rev.* **109**, 193 (1958).
[31] M. Gell-Mann and M. Levy, The axial vector current in beta decay, *Nuovo Cim.* **16**, 705 (1960).

[32] Y. Nambu, Axial vector current conservation in weak interactions, *Phys. Rev. Lett.* **4**, 380 (1960).
[33] M. L. Goldberger and S. B. Treiman, Decay of the pi meson, *Phys. Rev.* **110**, 1178 (1958).
[34] C. R. Sun and B. T. Wright, Radionuclide k^{37}, *Phys. Rev.* **109**, 109, (1958).
[35] S. Sakata, On a composite model for the new particles, *Prog. Theor. Phys.* **16**, 686 (1956).
[36] Y. Nambu and G. Jona-Lasinio, Dynamical model of elementary particles based on an analogy with superconductivity. *Phys. Rev.* **122**, 345 (1961).
[37] Y. Nambu and G. Jona-Lasinio, Dynamical model of elementary particles based on an analogy with superconductivity. *Phys. Rev.* **124**, 246 (1961).
[38] Y. Nambu and D. Lurie, Chirality conservation and soft pion production, *Phys. Rev.* **125**, 1429 (1962).

Chapter 11

Yoichiro Nambu: Visionary gentleman scientist

Lay Nam Chang

College of Science, Virginia Tech
Blacksburg, VA 24061, USA

My interactions with Yoichiro Nambu began in 1969, when he was at the peak of his creativity, and continued until his passage in 2015. I will describe below what I learned from his vast experience, his worldview, and in the process, how I also acquired a valued mentor and friend.

To see a World in a Grain of Sand
And a Heaven in a Wild Flower
Hold Infinity in the palm of your hand
And Eternity in an hour

William Blake — Auguries of Innocence

In many respects, extraordinary artists and scientists share many traits. They have exceptional creativity, of course, but they also have the ability to capture the world in unexpected and yet illuminating ways that imbue their creations with a life of their own. It is not possible to inquire how and whence the inspiration came. But it is possible to appreciate the sheer artistry they have given us. Professor Yoichiro Nambu belongs in this class of masters. I was fortunate to be working under Yoichiro in the late sixties and early seventies when his creativity was at peak levels. I was deeply influenced by his worldview on the unity of science, his unique way of approaching problems, and in the process, acquired a wise mentor and valued friend. What follows are my recollections of what for me were extremely exciting times.

I first met Professor Yoichiro Nambu in the winter of 1969, while interviewing for a job at the Enrico Fermi Institute (EFI). Y. C. Leung and I had just finished a paper on kinematic zeroes and singularities of helicity amplitudes involving massless particles, and their relationships to threshold behaviors, and to the requirements of gauge invariance [1]. We were wondering if better understandings along these lines would furnish us with another way to look at chiral symmetry breaking in strong interactions, with their attendant zeroes near threshold for soft pions.

I had looked forward very much to meeting Nambu. All through my graduate career, and during the two years I spent at MIT, Steve Weinberg always said that Nambu had a unique and likely ground breaking perspective on the idea of partial conservation of currents, and that if given a chance, there was much I could learn from him. I presented our results on the kinematic structure at the seminar, and awaited his verdict. He was extremely polite in his reaction, and suggested that while intriguing, our approach was probably incomplete when it came to chiral symmetry breaking. The basis for this intuition on his part is worth reviewing. In Yoichiro's mind, the Nambu–Goldstone theorem is only a part of chiral symmetry breaking. The equally important consequence for him was the existence of a scalar particle at twice the mass of the constituent fermions relative to which the symmetry is defined. For him, they were equally critical, and any approach that did not take both of them into consideration would be incomplete. But he urged us to continue to pursue the approach nonetheless. I subsequently found out that Yoichiro was ever the gentleman, and would always encourage everyone to pursue an idea to its logical conclusion.

I did get an offer from Chicago, and showed up at EFI that fall. At that time, theoretical physics was trying to understand the implications of the dual resonance model, and the multi-point Veneziano amplitude. While at MIT, I shared an office with Gabriele Veneziano, not knowing anything about that subject, except for the interesting suggestion of Lovelace on a variation that incorporated the soft pion zeroes [4]. I thought I had better study up on the field before coming to EFI. At MIT, Fubini and Veneziano had successfully factorized the n-point amplitude, revealing a rich spectrum that grows exponentially with mass, which is necessary to implement the idea of duality [2]. My attempts at generalizing the Lovelace amplitude to n-points were ultimately fruitless [5]. But just before I left for Chicago, someone handed me a preprint by Yoichiro, where he factorized the same n-point Veneziano amplitude in terms of a series of oscillators and coherent states [3].

When I met Yoichiro again, I asked him what he thought was the significance of the oscillators, and if they implied some continuum entity. He said two things in reply. First, he said they might represent a physical realization of the Dirac string for monopoles [6]. And second, they could be used to obtain a statistical description of high energy scattering, using the Kubo–Matsubara approach. In two meetings with him, he had opened up many different and intriguing approaches to pursue. And that would be the underlying theme of my subsequent interactions with him. Always willing to share ideas, and always also to listen.

Yoichiro, Peter Freund and I wrote a paper together exploring the statistical approach outlined above [7]. And then I decided to investigate the idea that strings underpinned the concept of duality in scattering amplitudes. I made it a point to see Yoichiro often to test out how that might work. He shared his time generously. Very early on, he felt that the two ends of an open string should be tied to monopoles that carry a triplet of internal quantum numbers. Because these Dirac-like strings represented objects that carried energy, and could also manifest themselves as closed entities, whatever gauge invariance that might be embedded in the system, should be broken. What that symmetry was and how it got broken became the crucial questions. But he did come up with what I thought was a good name for the quantum numbers. He labeled them D, N, and A. He gave a talk at the APS winter meeting held in Chicago in 1970, describing some of these ideas.

There are two things about Yoichiro's thinking that I took away from my interactions with him. He was a master craftsman, and a master magician. Ken Johnson once remarked to me that Nambu was the quintessential quantum mechanician. He understood quantum physics at a deep level, and had a profound feel for implications of gauge invariance. He was always looking at things from the canonical perspective of action principles and Hamiltonians. And at underlying local symmetries that might constrain the dynamics. The idea of the action that controlled strings therefore was very much at the forefront during our discussions. Through analogy with the action for a relativistic particle, which is simply the world line traced out by the system, the world sheet area, swept out by the string, was posited as the appropriate action [8]:

$$\begin{aligned}
\mathcal{S}_{\text{particle}} &= \int d\tau \sqrt{v^{\mu} v_{\mu}} \\
v^{\mu} &= \frac{dz^{\mu}}{d\tau} \\
\mathcal{S}_{\text{sheet}} &= \int d^2\zeta \sqrt{\sigma^{\mu\nu} \sigma_{\mu\nu}} \\
&= \int d^2\zeta \sqrt{g} \\
\sigma^{\mu\nu} &= \frac{\partial\{y^{\mu}\ y^{\nu}\}}{\partial\{\zeta^1\ \zeta^2\}}
\end{aligned}$$

Here z^{μ} refers to the position of the relativistic particle on the world line, τ is a parameter specifying a particular spot on that line, $\sigma^{\mu\nu}$ is the surface element on the world sheet, and g is the determinant of the metric on that sheet. y^{μ} are the positions of the string on the sheet, and ζ^a parametrize the coordinates on it. The question was what symmetry, broken or otherwise, would underpin the resultant dynamics.

In 1970, Freydoon (Fred) Mansouri joined EFI, and we shared an office. Pretty soon, he was part of the hunt for the underlying symmetry. We were still going back to the Dirac-string for inspiration, but soon gave that up, and focused instead on the underlying re-parametrization invariance of the action, and in particular the conformal symmetry. We very quickly came to the realization that the Virasoro algebra was a manifestation of the conformal symmetry, which was broken by the associated anomaly [9]. We ran our ideas by Yoichiro. To our great delight, he thought that we had the local symmetry we were looking for. He had been exploring alternative actions, and thought that what we had done could provide the right framework for these alternatives. One of these was in essence the "square" of the action we had been considering.

$$\begin{aligned} \mathcal{S} &= \int d^2\zeta \frac{1}{2} \partial_a y^\mu g^{ab} \partial_b y_\mu \\ &= \frac{1}{2} \mathrm{Tr}\, g \end{aligned}$$

We suggested writing up the results together. He insisted instead that we should do so on our own. The further modifications and clarifications would come later, he said, after our paper came out in the Physical Review [10]. They eventually appeared in a paper he wrote with Fred [11].

We took our time in writing up our results. This was because my term at EFI was ending in 1971, and I was looking for my next job. The idea of working on the theory of a relativistic string was quite bizarre at that time, and not conducive to getting a job offer. So I decided to stick to more conventional stuff, particle multiplicities in high energy scattering, for my job seminar. It wasn't until I got a job offer eventually that Fred and I submitted our paper for publication. Yoichiro was very patient, and appreciating my situation, never pressed us to hurry.

I left Chicago with deep personal regard for Yoichiro. He was the ultimate gentleman scientist, with a big heart. And a unique perspective on all aspects of physics. During my stay, he would ask me to do substitute teaching for him on many occasions. He would share with me his lecture notes on the topics he wanted covered. They were models of succinctness and originality, and characteristic of the way he expressed himself. Even on such a well-covered topic like the Fresnel relations governing reflection and refraction, he had a take that made the derivation that much more informative, through clever use of arguments based upon the underlying local symmetry in QED. I learned a lot as a result.

In my last days at EFI, the job situation in theoretical physics was abysmal. And I didn't have any offers for a long while. EFI had a policy of not keeping postdocs beyond two years. On his own, Yoichiro petitioned all members to grant an exception in my case, to act as a backstop, and succeeded in getting concurrence. He did this without telling me. I finally did get an offer to go to Penn. I will always remember the look of genuine joy when I informed him of this offer. It was only years later that I heard indirectly about the effort he made on my behalf. I was, and still am, very touched by this concern.

I tried to keep in contact with Yoichiro through the following years, wondering how he was, and what he was up to. Several times, I sought his advice on how to resolve difficulties I encountered, both professionally and personally. During our many conversations, we talked about numerous things, which extended well beyond physics. I always enjoyed these interactions, and felt I received good and wise counsel on many matters. He was generous with his time, and ever encouraging.

I believe it was Bruno Zumino who once remarked, "If you want to know what physics is like ten years from now, just find out what Nambu is doing now." Prescience touched by humility, grace and deep insight, that was the essence of Yoichiro Nambu. His influence on those lucky enough to have known and interacted with him is long-lasting and rewarding.

We will all miss him.

References

[1] L. N. Chang and Y. C. Leung, Kinematic structure for helicity amplitudes for massless particles, *Phys. Rev.* **185** (1969) 1945–1959.
[2] S. Fubini and G. Veneziano, Level Structures of dual resonance models, *Il Nuovo Cimento A* **64** (1969) 811–840.
[3] Y. Nambu, Quark Model and the factorization of the Veneziano amplitude, in *Conference on Symmetries*, Detroit, (1969). Fubini, Gordon

and Veneziano also suggested a similar use of the oscillators: S. Fubini, D. Gordon, G. Veneziano, A general treatment of factorization in dual resonance models, *Phys. Lett.* **B29** 679–682 (1969).

[4] C. Lovelace, A novel application of Regge trajectories, *Phys. Lett.* **B28** (1968) 264–268.

[5] A. P. Balachandran, L. N. Chang and P. Frampton, No go theorems for dual models, *Il Nuvo cimento* **A1** (1971) 545–552.

[6] P. A. M. Dirac, Quantized singularities in the electromagnetic field, *Proc. Royal Society* **A133** (1931) 60–72.

[7] L. N. Chang, P. Freund and Y. Nambu, Statistical approach to the Veneziano Model, *Phys. Rev. Lett.* **24** (1970) 628–631.

[8] Y. Nambu, Duality and hadrodynamics, in *Nambu: A Foreteller of Modern Physics*, edited by T. Eguchi and M. Y. Han, World Scientific Series in 20th Century Physics, Vol. 43, World Scientific, 2013.

[9] M. Virasoro, Subsidiary conditions and ghosts in dual resonance models, *Phys. Rev.* **D1** (1970) 2933–2936.

[10] L. N. Chang and F. Mansouri, Dynamics underlying duality and gauge invariance in dual resonance model, *Phys. Rev.* **D5** (1972) 2535–2542.

[11] F. Mansouri and Y. Nambu, Gauge conditions in dual resonance models, *Phys. Lett.* **B39** (1972) 375–378.

Chapter 12

The roots of QCD, Nambu's drama and humor

Moo-Young Han

Professor Emeritus of Physics
Duke University

The roots of quantum chromodynamics

The genesis of today's Quantum Chromodynamics, QCD, the sector of the Standard Model that governs the interaction among quarks and gluons, was clearly stated in page B1010 of our paper (Han and Nambu, 1965). We stated:

"We may characterize the hierarchy of interactions and their symmetries implied by the above model as follows.

First, the ***superstrong*** interactions responsible for forming baryons and mesons have the symmetry *SU(3)*$''$, and causes large mass splittings between different representations. The scale of mass involved would be comparable or larger compared to the baryon mass splittings, namely greater than 1 GeV.

The lowest state, i.e., *SU(3)*$''$ singlet states, would split according to *SU(3)*$'$, which would be the *SU(3)* group observed among the known baryons and mesons, with their ***strong*** interactions."

This is the first time in which a clear distinction was made between the interaction among the known hadrons and the interactions among quarks. The former was still called at that time as the *strong* interaction, as has been all along and we named the latter, the new interactions among quarks, the *superstrong* interaction. The terminologies have completely changed since then. Now, what we called the *superstong* interaction among quarks has become simply the strong interaction while the original *strong* interaction among the known hadrons disappeared completely, relegated to the 'molecular type residue force' among the constituent quarks of barons and mesons. This evolution in terminologies does not seem to be well understood.

As I have explained in my book, "From Photons to Higgs: A Story of Light (Second Edition)", Chapters 17 and 18, the $SU(3)''$ symmetry for quarks was modified in 1971 by Harald Fritzsch and Murray Gell-Mann as the exclusive symmetry for quarks, not to be related to any other interactions, neither electromagnetic nor weak. The three defining properties of $SU(3)''$ was named ***color*** charges, red, green and blue, carried by the quarks, conceptualized similar to the electric charges. And the symmetry group $SU(3)''$ was reincarnated as the $\boldsymbol{SU(3)_{\mathrm{color}}}$.

The $SU(3)''$ octet of gauge vector particles were named ***gluons*** of $SU(3)_{\mathrm{color}}$ and the QCD in its present form was launched by Fritzsch and Gell-Mann in 1971.

Nambu's drama and humor

At a personal level, we all fondly remember Nambu as a great friend who was kind, considerate, modest, soft-spoken and reserved in the finest tradition of the old Japanese way.

If Nambu is to be characterized in one short phrase, it is that he was a person of humble modesty and quiet dignity.

What is less widely known is that Nambu harbored a delightful penchant for drama on one hand and a well-hidden sense of humor on the other.

I was privileged to have personally experienced these two surprising sides of his personality.

The drama

The first occasion I witnessed his penchant for drama was in November 1974, when the news broke on the discovery of the J/ψ particle

Fig. 1. The original Yukawa Institute building.

simultaneously at the Brookhaven and Stanford laboratories — the so-called November Revolution.

In the fall of 1974, I spent a one-semester sabbatical leave at the Yukawa Institute of Kyoto University, formally called the Research Institute for Fundamental Physics, a forerunner of today's Research Institute for Theoretical and Experimental Physics.

My family and I were housed in a third floor apartment in the Konoe Hall, a short walking distance from the Institute.

Konoe Hall, specifically built for visiting foreign professors, was a four-story building but with a small footprint.

The first floor had a lounge and housed the caretaker's residence and each of the upper three floors had two apartments each, so six apartments in total.

One day in November, I stayed in the apartment instead of going to the Institute because it was a dreary day with light drizzle.

Sometime before noon, a staff member of the Institute phoned me to tell me that I have a telegram from Nambu that came that morning. It would be placed in my mailbox.

Fig. 2. Yukawa (1907–1981) in front of the Institute.

Fig. 3. In front of the Konoe Hall, in 1974.

I had never received a Western Union telegram from anyone before and this was from Nambu in Chicago. Quite alarmed and hoping it was not a bad news, I rushed to the Institute to open the telegram. It had strips of sentences from a telegraph machine pasted on it.

The first line of the telegram read, "Colored or charmed particle may have been discovered, period." Then it went on to describe that the particle was simultaneously discovered at BNL and SLAC and it had a mass around 3.1 GeV but with unexpectedly narrow width. Puzzled, and relieved that it was not an emergency, I showed it to Ziro Maki, then the Director of the Institute.

Maki immediately had a dozen copies of the telegram made (Remember this was 1974, so the Xerox machine had yet to be invented). The first copy was hand-delivered to Yukawa who happened to be in his office that day and others were distributed to permanent members of the Institute including Konuma and younger researchers that included Muta and several others whose names I cannot remember.

Well, all hell broke loose. Mass at 3.1 GeV and very narrow width?! People were first puzzled and then dazed, struggling to understand. We all congregated in the lobby and started shooting the breeze. Someone said (correctly) "Zweig–Iizuka Rule." Others interjected "a new quark?!" With only Nambu's telegram that did not quantify what was meant by narrow width, there wasn't much to go by and do — just wondering out loud.

Soon the word spread to every physics departments and research institutes all over Japan. The news caught Japanese physics community by storm. The telegram was not from some lowly Japanese physicists working in the United States. It was from Nambu, after all.

It is hard to imagine now, but in 1974, without the internet and email, that was how the entire Japanese physics community first got the news of the discovery of J/ψ particle, from the dramatic telegram from Nambu, to me, who happened to be in Kyoto at that time.

Nambu could have phoned either me or any of his colleagues in Japan or send an global express mail (no DSL nor FedEx yet). He certainly could have sent the telegram to Yukawa, Maki, Nishijima or others but he chose this melodramatic means of sending a rather cryptic Western Union telegram that while delivering exciting news, was much too brief for anyone to fully grasp the gravity of the situation.

Several Japanese physicists took the train to travel to the Institute and a bunch of people started working on the new particle, but other than its mass and the narrow width there was little to go on. By the next day many contacted their colleagues working in the United States and more news trickled in.

Either the next day or two (I don't remember exactly), Nambu followed up on his dramatic first telegram with the second telegram, just as dramatic. Now this is where the story gets interesting.

They did not phone me to say there was another telegram. Knowing what the issue was, they did not want to alarm me again. In any case I would be coming in soon and open the second telegram. For me, I was in no particular hurry.

There was a small take-out only Chinese restaurant a block away from Konoe Hall that cooked some heavenly tasting Chinese food and I decided to have an early lunch before leisurely walking to the Institute.

I learned later that my coming into office after lunch had driven many people at the Institute up the wall! Many people were pacing back and forth in front of my mailbox waiting for me to show up and open that damn second telegram from Nambu! They were getting pretty impatient but refrained from calling me again at my apartment. They even assigned someone to stand outside the front entrance and look down to let them know when I was spotted walking toward the Institute!

Even Yukawa came in early that day. I might have been the only person to make Yukawa actually wait! Yukawa was, after all, a semi-God to Japanese physicists!

The second telegram from Nambu contained the news of the discovery of the excited state of J/ψ particle, the J/ψ-prime, and its decay modes to the ground state in detail. Now, people had something to calculate. Never mind the explanation of the narrow width, just calculate.

By then many speculative ideas started floating around and one Japanese physicist whom I personally know published a short quick note to Nuovo Cimento Letters. The Letter referenced Nambu's telegram to me as a referemce. And that concerned me. What if some of early statements in the telegram turned out to be incomplete or even wrong?

At that point I felt the need to shield Nambu from the contents of his telegram becoming too public — at least not as a reference in a refereed journal. By the time the third telegram arrived with a little more details, I decided to share it only with Maki.

The three Western Union telegrams that Nambu chose to deliver the news in November 1974 were pure Nambu in his quiet penchant for a dramatic touch!

The humor

I was also able to glimpse the sudden explosion of Nambu's delicate sense of humor. It came completely unexpected, without any warning, in a warm balmy evening in Miami.

For about a dozen years starting in the mid-1960s, there was an important gathering of particle physicists on the campus of the University of Miami at Coral Gables, called the "Coral Gables Conference of High Energy Physics."

It was organized by Behram Kursunoglu, then the founding director of Center for Theoretical Physics at the University of Miami. Attendees were an admixture of two groups, a star-studded who's who of particle physics (any big names in the field were there) and carefully selected young researchers.

Kursunoglu was one of the earliest PhD students of Paul Dirac. Dirac had retired to the Florida State University in Tallahassee and as the 'honorary' member of the Institute he was always present at every conference. So were many 'permanent' participants that included Nambu.

After hashing out the cutting-edge state-of-the-art status of particle physics during the daytime, all participants were treated to a luxurious and specially arranged reception in the evening at

Fig. 4. Kursunoglu with Einstein.

Fig. 5. This was how Dirac actually looked during the conference period. Once he invited me to walk with him around a block or two and I was more than happy to oblige but he walked so brisk and fast I had trouble keeping up with his pace!

luxurious resort settings. Kursunoglu had this great skill — always finding a millionaire to host lavish dinner reception every year. It was one of the highlights of the conference, which was always held at a top venue. Some years, it was a garden reception at a bayside lawn of a mansion, complete with a string quartet, or at a floating resort club house. You get the picture.

Nambu's sense of humor came out at one of these dinner receptions one year (I don't remember which year). After being wined and dined at a sumptuous reception held at a country club on the Biscayne seafront, all participants strolled out and were walking slowly toward the chartered bus that would take us back to our hotels. It was a balmy evening with warm tropical breeze and we were all taking our time slowly walking toward the bus.

Nambu was a few steps behind me and he came up to me and we walked together enjoying the evening breeze. Then he asked me a question and our conversation went like this:

Nambu, "Do you know how our host this evening made his fortune?"

Han, "No, I sure don't."

Nambu, "He made his fortune by making Sakku!"

It took a second or two for this hilarious irony to sink into my head. I started laughing. Nambu started laughing.

Sakku is a Japanized English word meaning Sack and it is the colloquial word for male condom.

We laughed some more and before we knew we hit a resonance of laughs and there was no holding back. We just lost it completely. Totally oblivious to other participants alongside of us, we just let out hearty roar of laughs. We could not stop.

We stopped walking and just stood there laughing our heads off, like two crazy people.

I never heard Nambu laugh so hard, so loud and for so long. Sometimes I can still hear his laughs. That was one moment Nambu lost his control and let his sense of humor explode out.

Chapter 13

Hommage à Nambu

Pierre Ramond

Institute for Fundamental Theory, Department of Physics
University of Florida, Gainesville, FL 32611, USA
ramond@phys.ufl.edu

Reminiscences of my intellectual and personal interactions with Professor Nambu, and discussion of his contributions to string theory. Switching to my own research, by expressing the string coordinates as bispinors, I suggest a kinematical framework for the interacting (2, 0) theory on the light-cone by using chiral constrained "Viking" superfields.

1. Nambu Sensei

In 1969 freed from the Mandelstam triangle with a PhD, I started studying Veneziano's model that summer in Trieste under the aegis of Jean Nuyts and Hirotaka Sugawara. I was focused on equations although everything was cast in the language of amplitudes. I arrived in Weston, near Chicago, the site of the National Accelerator Laboratory (today's FermiLab), and started collaborating with fellow postdoc Lou Clavelli. When I mentioned that the spectrum of masses extracted from the amplitudes reminded me of something I had seen before, Lou said simply "*Nambu says it is a string!*", and added, "*would you like to meet him?*"

Lou, a former student of Nambu, arranged a meeting for lunch at the quadrangle Club. I was not sure what to expect. At NAL we were tasked by Bob Wilson to entertain the senior physicists who came by to inspect the circular hole in the ground; most looked through us and never showed interest in our work, or so I felt. To my delight, the gentleman at the table made us feel at ease, expressed interest, and volunteered to guide us in our work since there were no senior theorists at NAL. He even paid for the lunch! This cemented a lifelong friendship and deep respect from this writer. Nambu, a genius at making connections and reasoning by analogy, acted like a postdoc! There were many subsequent meetings at which we explained our progress and sought his advice. Not once was the experience negative.

In the fall of 1970, I went to Princeton where my wife was attending a training course; Nambu was on sabbatical at the Institute and I told him about the fermion equation with some trepidation. He was totally encouraging. His approval sustained me since NAL had just fired us after two years of a promised three-year position! I have since always asked his advice when confronted with a new idea or making a decision. When I was considering a move from Caltech to the University of Florida, he mentioned his own move with Nishijima from Tokyo to Osaka, and essentially said that with the right people, nothing else mattered.

In 1975 Paris, Nambu and I went to listen to Marie Claire Alain at the Grandes Orgues de Saint Sulpice, which I had advertised as a treat. On the way I found out that he spoke some french, and that his approach to physics was motivated by instinct and the joy of discovery, consistent with his presence in a young person's subject (US senior theorists were nowhere to be seen). Sadly, Mme Alain was not practicing that day!

In 1998 at Trieste, I was giving a talk on neutrino oscillations. Sandip Pakvasa had told me of the seminal role of the Nagoya school on neutrino mixing, and I called the lepton mixing matrix, the MNS matrix, after Maki, Nakagawa and Sakata who had invented it. After the talk, I was surrounded by physicists irate that I had not mentioned Pontecorvo, who had introduced the idea of neutrino–antineutrino oscillations in analogy with K–$\bar{K}$ mixing. But it was worth it: Nambu came over, shook my hand and just said, "*thank you*".

As a final vignette, Professor Nambu gave me the ultimate compliment by attending my sixtieth birthday. I will always cherish the lessons he taught me, in physics and beyond.

2. The humble genius

With a degree in electrical engineering in 1965, my desire in life was to study theoretical physics, driven by an ill-defined fascination with the elegance of nature. Accepted by Syracuse University to study General Relativity, I was soon "turned" into a particle physicist by Alan McFarlane's course on Advanced Quantum Mechanics. I became a student of Professor E.C.G. Sudarshan who, before leaving for India on a sabbatical, asked me to learn all about infinite component wave equations, and directed me to a paper by Yoichiro Nambu [1]; thus began my intellectual travels with Professor Nambu.

It was for me an arduous path to travel, but the vistas were worth the trip, even though it took me some time to appreciate them. I learned from this paper that symmetries are crucial, and it reinforced my feeling that equations could be "beautiful" even for complicated systems, in particular Majorana's equation [2].

Majorana, unhappy with Dirac's negative energy solutions [3], had modified it to

$$[E + \vec{\alpha} \cdot \vec{p} + \beta M]\Psi = 0,$$

where β is strictly positive. The solutions arranged themselves into unitary representations of the Lorentz group, with an infinite number of particles with spin j and masses $M_j = \frac{2M}{2j+1}$. With the discovery of the positron, Majorana's work was forgotten.

It is obvious by inference that Nambu had invented Majorana's equation. In his 1966 paper, he says "*Even equations with an infinite number of levels have also been studied in the past, including the little known but remarkable work by Majorana in 1932*", and "*... thanks Prof. F. Gürsey and Mr. J. Cronin for first calling his attention to these papers*". Such was Nambu's elegance and fairness.

A little background on Veneziano's Dual Resonance Model and its morphing into Strings as it pertains to Nambu's impact on me and my generation:

- In August 1967, R. Dolen, D. Horn and C. Schmid [4] discovered a surprising feature of the $\pi - N$ amplitude. In the fermionic s-channel, it is dominated by the Δ resonances, with a train of Breit–Wigner shapes. In the bosonic t-channel, the same amplitude is dominated by the ρ meson. In the Regge picture of the day, ρ exchange contributes to the s-channel a continuous curve, which traces the average of the Δ resonances. It suggests a duality between the s and t channels and a new input to the bootstrap program. The hunt was on for amplitudes with s and t channel duality.
- In July 1968, G. Veneziano [5] proposed the amplitude.
- At the Wayne State meeting in June 1969, Nambu remarked that the states that mediate Veneziano amplitude are due to quantum strings [6].
- In December 1969, Virasoro [7] found an infinite number of decoupling conditions of the negative norm states extracted through factorization of the Veneziano amplitudes. The cost is high: a tachyon, and a massless vector boson unexpected in strong interaction physics.

2.1. *Dual model of hadrons*

At the American Physical Society Meeting in January 1970 in Chicago, Nambu incorporated into the Veneziano model two vistas, Harari–Rosner quark diagrams, and the "DNA" Han–Nambu model. He believed that the Veneziano model provided a dynamical scheme to hold the Han–Nambu units (D, N, and A) together, with the string equation at the central focus of the new dynamics.

Nambu turned next to Virasoro's decoupling conditions, and in a spectacular footnote, identified their origin: "*These transformations may be expressed in terms of the energy momentum tensor $T_{\alpha\beta}$ $(\alpha, \beta = 0, 1)$ in the internal space (η, ξ)... The set of $T_{\alpha\beta}[f_n]$'s generate a Lie algebra...*" In his mind, these decoupling conditions were generated by the conformal symmetry of two-dimensional system. The importance of this remark was not truly appreciated at the time, but there is more!

In another leap, he brought in Dirac's magnetic monopole where the magnetic flux flows along a string. He said "*... since we have the string, why not take advantage of it, and add Dirac's electromagnetic Lagrangian to ours!... quarks at the end of a meson string will be assigned equal and opposite magnetic charges, so there will be a very strong magnetic interaction between them...*" In this talk, Nambu, ever the magician, brought in analogies and insight in the description of the new subject of Veneziano amplitudes!

2.2. *Duality and hadrodynamics*

Six months later, we got a more detailed peek at Nambu's thinking, encompassed in this lecture at the Copenhagen Summer Symposium [6]. Actually, Nambu never delivered it. He and his son John were passed by an obnoxious driver, and Nambu decided to overtake him in turn and gunned his car! This resulted in blown gaskets and a week in Wendover at the Nevada–Utah border!

His primary concern was in "*... guessing at the dynamics of hadrons which underlie the Veneziano model*". In spite of nice features like linearly rising Regge trajectories, he was concerned by the ghosts and tachyon in the model, and with some humour asked "*How many ghosts are real. This is the most serious question of principle that haunts us, especially us the theorists.*"

After reviewing the point particle description he proposed the famous equation,

$$I = \int |d\sigma_{\mu\nu} d\sigma^{\mu\nu}|^{\frac{1}{2}} = \iint |2 \det g|^{\frac{1}{2}} d^2\xi;$$

together with the energy-momentum tensor,

$$T_{\alpha\beta} = \frac{1}{2\pi}\left(g_{\alpha\beta} - \frac{1}{2} g_{\alpha\beta} g_{\gamma\delta} g^{\gamma\delta}\right),$$

identifying the Virasoro conditions as its moments

$$L_n^{\pm} = \int_0^{\pi} (T_{00} \pm T_{01})(\xi) e^{2in\xi} d\xi$$

and the Virasoro algebra (no c-number)

$$[L_n^{\pm}, L_m^{\pm}] = 2(n-m)L_{n+m}^{\pm}; \quad [L_n^{\pm}, L_m^{\mp}] = 0.$$

An insightful remark followed: "*... finite dimensional Lorentz tensors lead to negative probabilities; infinite dimensional representations lead to tachyons.*" Which to choose, ghosts or tachyons, "*... the agony of making the choice ...*" Formulating everything in terms of Lorentz oscillators leads to Gaussian (bad) form factors, but infinite component unitary representations have a tachyon.

There followed a string of insights and analogies for the dynamical structure of hadrons:

— Consistency of the Harari–Rosner quark diagrams with the string picture suggested a dynamical models for strings, as the glue that bound linear molecules of the constituents. Interactions are simply the breaking of one chain into two. The constituents were the Han–Nambu particles with $SU(3)''$ (today's color), with singlet observables.
— Strings could be viewed as a continuous set of $T\bar{T}$ states, and one should think of them as two-dimensional antiferromagnet, suggesting Onsager's solution of the Ising model as a prototype.
— He considered adding Dirac monopoles to the string picture with dyons at the end of the strings. The large P and T violations worried him because the electric dipole of the neutron, if it exists at all, is so very small.
— Nambu proposed a statistical approximation based on the large number of states in the Veneziano model. He suggested a grand canonical ensemble of resonances, where the strings interacted with one another, from which the Pomeron emerged.
— Concerned with the SLAC-MIT experiments which showed power and not Gaussian form factors, Nambu tried to remedy the situation by adding a fifth dimension which would generate a flat Regge trajectory and "explained" the data.

This was my poor attempt to describe this extraordinary display of physics, originality and computational power.

Every physics student should be asked to read "Duality and hadrodynamics" [6], for the diversity of thoughts and for the display of Nambu's genius.

2.3. *Magnetic and electric confinement of quarks*

At a Paris meeting in June 1975, "Extended Systems in Field Theory", Nambu sought yet again an underlying theory of quarks that reproduced strings, combining monopoles with London's phenomenological theory of superconductivity ($j_\mu \sim A_\mu$), leading to an Abelian theory with magnetic monopoles at the end of open strings. But, generalizing to a non-Abelian magnetic flux octet density $\rho^i(x)$ at each point of the string, Eguchi showed consistency only for the two Abelian $SU(3)$ generators. There appeared to be no such non-Abelian picture.

He then addressed electric confinement using the B-field of string theory [8]. There strings are like vortices, and the B-field represents a particle with zero-helicity. He ended up with a phenomenological Lagrangian and a London-like equation, equating the antisymmetric B-field with the string current. Again there was no non-Abelian generalization. In passing he generalized the Abelian invariance of the two-form to higher forms.

3. Eight bosons and eight fermions

In this section begins my scientific contribution to this volume in honor of Professor Nambu.

Forty years later, the B-field is at the heart of the six-dimensional $(2,0)$ superconformal theory. It is the lack of a non-Abelian generalization of its conformally invariant coupling to closed string currents that stands in the way of its formulation.

The kinematics of superconformal theories with eight bosons and eight fermions is particularly simple on the light-cone, where $x^{\pm} = (x^0 \pm x^1)/\sqrt{2}$, $\partial^{\pm} = (\partial^0 \pm \partial^1)/\sqrt{2}$, together with transverse coordinates. With four complex Grassmann variables θ^m and $\bar{\theta}_m$, the chiral derivatives,

$$d^m = -\frac{\partial}{\partial\bar{\theta}_m} - \frac{i}{\sqrt{2}}\,\theta^m\,\partial^+, \quad \bar{d}_n = \frac{\partial}{\partial\theta_n} + \frac{i}{\sqrt{2}}\,\bar{\theta}_n\,\partial^+,$$

satisfy

$$\{d^m, \bar{d}_n\} = -i\sqrt{2}\,\delta^m_n\,\partial^+,$$

and the eight bosonic and eight fermionic degrees of freedom are assembled into the "Viking" superfield [9],

$$\Phi(y) = \frac{1}{\partial^+} A(y) + \frac{i}{\sqrt{2}}\,\theta^{mn}\,\overline{C}_{mn}(y) + \frac{1}{12}\,\theta^{mnpq}\,\epsilon_{mnpq}\,\partial^+ \bar{A}(y)$$
$$+ \frac{i}{\partial^+}\,\theta^m\,\bar{\chi}_m(y) + \frac{\sqrt{2}}{6}\,\theta^{mnp}\,\epsilon_{mnpq}\,\chi^q(y); \tag{1}$$

it is chiral,

$$d^m\,\Phi(y) = 0, \quad \text{if } y = x^- - i\frac{\theta^m\bar{\theta}_m}{\sqrt{2}},$$

and obeys the "inside-out" constraint,

$$\bar{d}_m\,\bar{d}_n\,\Phi = \frac{1}{2}\,\epsilon_{mnpq}\,d^p\,d^q\,\overline{\Phi}. \tag{2}$$

The additional operators,

$$q^m = -\frac{\partial}{\partial\bar{\theta}_m} + \frac{i}{\sqrt{2}}\,\theta^m\,\partial^+, \quad \bar{q}_n = \frac{\partial}{\partial\theta^n} - \frac{i}{\sqrt{2}}\,\bar{\theta}_n\,\partial^+,$$

anticommute with the chiral derivatives,

$$\{q^m,\, \bar{d}_n\} = \{q^m,\, d^n\} = 0,$$

and satisfy

$$\{q^m,\, \bar{q}_n\} = i\sqrt{2}\,\delta^m_n\,\partial^+.$$

They generate the linear action of $SO(8)$ on Φ in terms of its $SO(6)\times SO(2) = SU(4)\times U(1)$ subgroup, with parameters ω^m_n and ω,

$$\delta_{SO(6)}\,\Phi = i\omega^m{}_n\left(q^n\,\bar{q}_m - \frac{1}{4}\delta^n_m\,q^l\,\bar{q}_l\right)\frac{1}{\partial^+}\,\Phi,$$

$$\delta_{SO(2)}\,\Phi = \frac{i\omega}{4}\,(q^m\,\bar{q}_m - \bar{q}_m\,q^m)\,\frac{1}{\partial^+}\,\Phi. \tag{3}$$

The $SO(8)/SO(6)\times SO(2)$ coset transformations,

$$\delta_{\overline{coset}}\,\Phi = i\omega^{mn}\,\bar{q}_m\,\bar{q}_n\,\frac{1}{\partial^+}\,\Phi, \quad \delta_{coset}\,\Phi = i\omega_{mn}\,q^m\,q^n\,\frac{1}{\partial^+}\,\Phi, \tag{4}$$

make up the twenty-eight $SO(8)$ transformations.

3.1. *Four dimensions*

This superfield was first used [9] to describe the $\mathcal{N}=4$ Super-Yang–Mills theory in $D=4$ dimensions. In the decomposition

$$SO(8)\supset SO(2)\times SO(6),$$

$SO(2)$ is identified with the transverse little group, and $SO(6)=SU(4)$ is the $\mathcal{R}$-symmetry. The constrained chiral superfield now carries an adjoint index for the gauge group, and its components are identified with the canonical physical degrees of freedom of the theory,

$$\Phi\longrightarrow\varphi^{C}(y,x,\bar{x}),$$

where x and its conjugate are the transverse coordinates. The kinematical supersymmetries Q^n and $\bar{Q}_n$, are q^m and $\bar{q}_n$, acting linearly on the superfields,

$$\delta_{\bar{\epsilon}_n Q^n}\varphi^C=\bar{\epsilon}_n q^n\varphi^C,\qquad \delta_{\epsilon^n\bar{Q}_n}\varphi^C=\epsilon^n\bar{q}_n\varphi^C.$$

The large superconformal symmetry, $PSU(2,2|4)$ of the $\mathcal{N}=4$ theory fully determines its dynamics. This uniqueness is summarized by the dynamical supersymmetries $\boldsymbol{\mathcal{Q}}^n$ and $\bar{\boldsymbol{\mathcal{Q}}}_n$ transformations [10],

$$\delta^{SYM}_{\varepsilon\overline{\boldsymbol{\mathcal{Q}}}}\,\varphi^A=\frac{1}{\partial^+}\Big((\bar{\partial}\,\delta^{AB}-gf^{ABC}\partial^+\varphi^B)\varepsilon^m\bar{q}_m\varphi^C\Big),\tag{5}$$

where A,B,C are adjoint indices of the gauge group. It generates by commutation a non-linear realization of the $PSU(2,2|4)$ symmetry. Split into a "kinetic term" with one transverse derivative, and the non-linear "interaction term" without transverse derivative, it allows for a perturbative description. It was derived solely from symmetries, and not from a Lagrangian (although it could have been).

3.2. *Six dimensions*

Let us apply a similar approach to the $(2,0)$ theory in six dimensions with $OSp(6,2|4)$ symmetry [11], with a lopsided assignment of the spin little group of eight bosons and eight fermions.

Although it is a fully interacting theory [12], the variables responsible for a non-linear representation of $OSp(6,2|4)$ have not been determined.

Unlike the $\mathcal{N}=4$ theory, the kinematics of the (2,0) demands (see later) the absence of a perturbative limit, robbing us an important tool.

Assuming eight bosons and eight fermions, a description in terms of Viking superfields Φ is expected from the kinematical constraints imposed by the symmetries, although the physical meaning of the superfield components is unknown (primary operators, hidden degrees of freedom?). In the free (2,0) theory, the components are canonical local fields.

Our approach is to find the best kinematical framework to narrow down, through symmetries, the non-linear representation.

The transverse little group $SO(4)\sim SU(2)_L\times SU(2)_R$ splits into two parts. The spin part of $SU(2)_L$ is extracted from the $SO(8)$ transformations (3–4), through the decomposition,

$$SO(8)\supset SU(2)_{spin}\times Sp(4)_{\mathcal{R}}, \tag{6}$$

where $Sp(4)\sim SO(5)$ are the $\mathcal{R}$-symmetries. The eight boson degrees of freedom split into an $\mathcal{R}$-quintet of scalar fields and an $\mathcal{R}$-singlet spin triplet corresponding to a self-dual second rank antisymmetric tensor in the transverse directions; the eight fermions are spinors under both $SU(2)_L$ and $SO(5)$.

$$\mathbf{8}_b=(\mathbf{3},\,\mathbf{1})+(\mathbf{1},\,\mathbf{5}),\quad \mathbf{8}_f=(\mathbf{2},\,\mathbf{4}). \tag{7}$$

3.2.1. *$\mathcal{R}$-symmetry*

The $\mathcal{R}$-symmetry action on the superfield is linear,

$$\delta_{SO(5)}\Phi=\frac{i}{2\sqrt{2}\partial^+}\alpha^{mn}(\boldsymbol{C}_{mp}q^p\overline{q}_n+\boldsymbol{C}_{np}q^p\overline{q}_m)\,\Phi=\alpha^{mn}q\,T_{mn}\,\overline{q}\frac{1}{\partial^+}\Phi, \tag{8}$$

where $\alpha^{mn}=\alpha^{nm}$ are the ten $SO(5)$ rotation angles, and

$$T^{ab}=-\frac{i}{2}\Gamma^a\Gamma^b,\quad a\neq b,$$

are their generators, where the five (4×4) Dirac matrices form a Clifford algebra. The charge conjugation matrix satisfies,

$$\mathcal{C} = \mathcal{C}^{-1} = -\mathcal{C}^T, \quad (\mathcal{C}\Gamma^a)^T = -(\mathcal{C}\Gamma^a), \quad \mathcal{C}\Gamma^a\mathcal{C}^{-1} = (\Gamma^a)^T,$$

and serves as a metric: matrices with lower indices are defined by

$$\mathcal{C}^{mn} \equiv (\mathcal{C}_{mn})^* = \frac{1}{2}\epsilon^{mnpq}\mathcal{C}_{pq}, \quad \mathcal{C}_{mn}\mathcal{C}^{np} = \delta_m^p.$$

3.2.2. *Transverse little group*

The transverse coordinate matrix,

$$X = \begin{pmatrix} x & -x' \\ \bar{x}' & \bar{x} \end{pmatrix} \sim (\mathbf{2}, \mathbf{2}), \tag{9}$$

is labelled by the dimensions of the $SU(2)_L \times SU(2)_R$ representations. The transverse derivatives, ∂, $\overline{\partial}$, ∂', and $\bar{\partial}'$, which satisfy $\partial\,\overline{x} = \overline{\partial}\,x = \partial'\,\overline{x}' = \overline{\partial}'\,x' = 1$, are assembled into a transverse derivative matrix that has the same transformation properties,

$$\nabla = \begin{pmatrix} \partial & \partial' \\ \bar{\partial}' & -\bar{\partial} \end{pmatrix} \sim (\mathbf{2}, \mathbf{2}).$$

The action of $(SU(2)_L \times SU(2)_R)$ on the coordinates is,

$$X \to X' = \mathcal{U}_L\, X\, \mathcal{U}_R^\dagger, \quad \mathcal{U}_{L,R} = \exp\left(\frac{i}{2}\,\vec{\omega}_{L,R} \cdot \vec{\tau}\right),$$

with $SU(2)_L$ generators,

$$L_+^{orb} = x\partial' - x'\partial, \quad L_-^{orb} = \bar{x}'\bar{\partial} - \bar{x}\bar{\partial}', \quad L_3^{orb} = \frac{1}{2}(x\bar{\partial} - \bar{x}\partial + x'\bar{\partial}' - \bar{x}'\partial').$$

Similarly for the $SU(2)_R$ generators,

$$R_+^{orb} = \bar{x}'\partial - x\bar{\partial}', \quad R_-^{orb} = \bar{x}\partial' - x'\bar{\partial}, \quad R_3^{orb} = \frac{1}{2}(x\bar{\partial} - \bar{x}\partial - x'\bar{\partial}' + \bar{x}'\partial').$$

Together they form the orbital part of the light-cone little group. On the other hand, spin is added asymmetrically,

$$SU(2)_L : \vec{L} = \vec{L}^{orb} + \vec{S}, \quad SU(2)_R : \vec{R} = \vec{R}^{orb}, \tag{10}$$

with the spin part written in terms of q^n and $\bar{q}_n$,

$$S_+ = \frac{i}{2\sqrt{2}}\frac{1}{\partial^+}\,\bar{q}\,\boldsymbol{C}\,\bar{q}, \quad S_- = \frac{i}{2\sqrt{2}}\frac{1}{\partial^+}q\,\boldsymbol{C}\,q,$$

$$S_3 = \frac{i}{4\sqrt{2}\partial^+}\,(q^m\,\bar{q}_m - \bar{q}_m\,q^m\,).$$

The superfield transforms under the transverse little group as,

$$\delta_{\vec{\omega}_L\cdot\vec{L}}\,\Phi = \frac{i}{2}\vec{\omega}_L\cdot\vec{L}\,\Phi, \quad \delta_{\vec{\omega}_R\cdot\vec{R}}\,\Phi = \frac{i}{2}\vec{\omega}_R\cdot\vec{R}\,\Phi, \tag{11}$$

so that $SU(2)_R$ is purely orbital. The kinematical supersymmetries,

$$\overline{Q}^{\alpha}_{m} = \begin{pmatrix} \bar{q}_m \\ (\boldsymbol{C}q)_m \end{pmatrix}, \quad Q^{m\alpha} = \begin{pmatrix} (\boldsymbol{C}\bar{q})^m \\ q^m \end{pmatrix} \sim (\mathbf{2},\mathbf{1}),\ \alpha = 1,2,$$

are $SO(5)$ spinors quartets, $SU(2)_L$ doublets, and $SU(2)_R$ singlets, which satisfy

$$\{\overline{Q}^{\alpha}_{m},\, Q^{n\beta}\} = i\sqrt{2}\epsilon^{\alpha\beta}\delta^n_m\partial^+,$$

and act linearly on the superfields,

$$\delta^{kin}_{\varepsilon\overline{Q}}\Phi \equiv \varepsilon^{m}_{\alpha}\overline{Q}^{\alpha}_{m}\Phi, \quad \delta^{kin}_{\bar{\varepsilon}Q}\Phi \equiv \bar{\varepsilon}_{m\alpha}Q^{m\alpha}\Phi. \tag{12}$$

The kinematical conformal supersymmetries are $SO(5)$ spinors with opposite chirality, $SU(2)_L$-singlets and $SU(2)_R$-doublets,

$$\overline{S}^{\dot{\alpha}}_{m} = -i\begin{pmatrix} \bar{q}_m\bar{x}' + (q\boldsymbol{C})_m x \\ \bar{q}_m\bar{x} - (q\boldsymbol{C})_m x' \end{pmatrix},$$

$$S^{m\dot{\alpha}} = -i\begin{pmatrix} q^m x + (\bar{q}\boldsymbol{C})^m\bar{x}' \\ -q^m x' + (\bar{q}\boldsymbol{C})^m\bar{x} \end{pmatrix} \sim (\mathbf{1},\mathbf{2}),$$

and satisfy,

$$\{\overline{S}^{\dot{\alpha}}_{m},\, S^{n\dot{\beta}}\} = 2i\sqrt{2}\delta^n_m\epsilon^{\dot{\alpha}\dot{\beta}}(x\bar{x} + x'\bar{x}')\partial^+.$$

They also act linearly on the superfields,

$$\delta^{kin}_{\varepsilon\overline{S}}\Phi = \varepsilon^{m}_{\dot{\alpha}}\overline{S}^{\dot{\alpha}}_{m}\Phi, \quad \delta^{kin}_{\bar{\varepsilon}S}\Phi = \bar{\varepsilon}_{m\dot{\alpha}}S^{m\dot{\alpha}}\Phi. \tag{13}$$

Their commutators with the kinematical supersymmetries yield transverse boosts which transform as $(\mathbf{2},\mathbf{2})$. For future reference, note that the only translation invariant $SU(2)_R$-doublets contain only transverse derivatives,

$$\begin{pmatrix} \partial \\ -\partial' \end{pmatrix}, \quad \begin{pmatrix} \bar{\partial}' \\ \bar{\partial} \end{pmatrix} \sim (\mathbf{1}, \mathbf{2}). \tag{14}$$

3.2.3. *Kinematical constraints*

The dynamical supersymmetries, denoted here by the bold face variations $\boldsymbol{\delta}^{(2,0)}_{\varepsilon\overline{\mathcal{Q}}}$, acting on any Viking superfield, must satisfy a number of kinematical constraints:

- $\boldsymbol{\delta}^{(2,0)}_{\varepsilon\overline{\mathcal{Q}}}$ preserves chirality,
- $\boldsymbol{\delta}^{(2,0)}_{\varepsilon\overline{\mathcal{Q}}}$ is translationally invariant (does not generate explicit transverse coordinates).
- $\boldsymbol{\delta}^{(2,0)}_{\varepsilon\overline{\mathcal{Q}}}$ rotates under $SU(2)_R$ as:

$$\left[\delta_{SU(2)_R}, \boldsymbol{\delta}^{(2,0)}_{\varepsilon\overline{\mathcal{Q}}}\right] = \boldsymbol{\delta}^{(2,0)}_{\varepsilon'\overline{\mathcal{Q}}}.$$

  Anticommutators of kinematical and dynamical supersymmetries are transverse space translations which transform as $(\mathbf{2}, \mathbf{2})$ under the little group. Kinematical supersymmetries transform as $(\mathbf{2}, \mathbf{1})$, so that $\boldsymbol{\delta}^{(2,0)}_{\varepsilon\overline{\mathcal{Q}}}$ transforms as an $SU(2)_R$ spinor. Since the only (translationally invariant) $SU(2)_R$ spinors are given by Eq. (14), the action of $\boldsymbol{\delta}^{(2,0)}_{\varepsilon\overline{\mathcal{Q}}}$ induces terms with an odd number of transverse derivatives. All terms have an odd number of transverse derivatives; this is the fundamental difference with $\mathcal{N} = 4$-SYM. The light-cone Hamiltonian contains only terms with transverse derivatives: *Kinematics alone implies that the* (2,0) *theory has no perturbative limit.*
- $\boldsymbol{\delta}^{(2,0)}_{\bar{\varepsilon}\mathcal{Q}}$ and $\boldsymbol{\delta}^{(2,0)}_{\varepsilon\overline{\mathcal{Q}}}$ are $SU(2)_L$-invariant:

$$\left[\delta_{SU(2)_L}, \boldsymbol{\delta}^{(2,0)}_{\varepsilon\overline{\mathcal{Q}}}\right] = 0.$$

- $\boldsymbol{\delta}^{(2,0)}_{\bar{\varepsilon}\boldsymbol{\mathcal{Q}}}$ and $\boldsymbol{\delta}^{(2,0)}_{\varepsilon\overline{\boldsymbol{\mathcal{Q}}}}$ are $SO(5)$ spinors:

$$\left[\delta_{SO(5)},\, \boldsymbol{\delta}^{(2,0)}_{\bar{\varepsilon}\boldsymbol{\mathcal{Q}}}\right] = \boldsymbol{\delta}^{(2,0)}_{\bar{\varepsilon}'\boldsymbol{\mathcal{Q}}}, \quad \left[\delta_{SO(5)},\, \boldsymbol{\delta}^{(2,0)}_{\varepsilon\overline{\boldsymbol{\mathcal{Q}}}}\right] = \boldsymbol{\delta}^{(2,0)}_{\varepsilon'\overline{\boldsymbol{\mathcal{Q}}}},$$

with $\bar{\varepsilon}'_m = 2\alpha_{mn}\bar{\varepsilon}_n$, and $\varepsilon'^m = 2\alpha^{mn}\varepsilon^n$s.
- The commutator with the Lorentz generator J^+ yields a kinematical operation:

$$\left[\delta_{J^+},\, \boldsymbol{\delta}^{(2,0)}_{\bar{\varepsilon}\boldsymbol{\mathcal{Q}}}\right] = \delta_{\bar{\varepsilon}Q} = \bar{\varepsilon}Q.$$

Since $J^+ = ix\partial^+$ is itself kinematical, the transverse coordinate disappears by commutation: $\boldsymbol{\delta}^{(2,0)}_{\bar{\varepsilon}Q}$ must contain at least one term linear in transverse derivatives.
- The same conclusion obtains by considering the right-hand-side of the commutator

$$\left[\delta_{K^+},\, \boldsymbol{\delta}^{(2,0)}_{\bar{\varepsilon}\boldsymbol{\mathcal{Q}}}\right] = \delta_{\bar{\varepsilon}S},$$

using $K^+ = i(x\bar{x} + x'\bar{x}')\partial^+$.
- The engineering dimensions of $\boldsymbol{\delta}^{(2,0)}_{\bar{\varepsilon}\boldsymbol{\mathcal{Q}}}$ and $\boldsymbol{\delta}^{(2,0)}_{\varepsilon\overline{\boldsymbol{\mathcal{Q}}}}$ fix the relative number of ∂^+ derivatives.
- The operator

$$\Delta = D - J^{+-} = x_i\partial_i + \Delta_\phi,$$

where Δ_ϕ is the conformal dimension, has simple transformation properties,

$$\left[\delta_\Delta,\, \boldsymbol{\delta}^{(2,0)}_{\bar{\varepsilon}\boldsymbol{\mathcal{Q}}}\right] = i\boldsymbol{\delta}^{(2,0)}_{\bar{\varepsilon}\boldsymbol{\mathcal{Q}}}, \quad \left[\delta_\Delta,\, \boldsymbol{\delta}^{(2,0)}_{\varepsilon\overline{\boldsymbol{\mathcal{Q}}}}\right] = i\boldsymbol{\delta}^{(2,0)}_{\varepsilon\overline{\boldsymbol{\mathcal{Q}}}}.$$

If the action of $\boldsymbol{\delta}^{(2,0)}_{\bar{\varepsilon}\boldsymbol{\mathcal{Q}}}$ induces n_∂ transverse derivatives, and n_i local operators $\mathcal{O}_i$ with $\Delta_{\mathcal{O}_i}$,

$$n_\partial = 3 - \sum_i n_i \Delta_{\mathcal{O}_i}.$$

A canonical superfield with $\Delta_\phi = 2$ describes the free case with $n_\partial = 1$. The general case implies either non-locality in the transverse space or negative conformal dimensions.

3.2.4. *Free theory: The tensor multiplet*

Further progress requires the identification of the components in the chiral superfield. If the components are canonical local fields, the decomposition (7) yields the tensor supermultiplet; its bosons split into a transverse self-dual two-form, and a quintet of scalars.

The kinematical requirements discussed above lead to a unique linear implementation

$$\boldsymbol{\delta}^{(2,0)}_{\varepsilon\overline{\mathcal{Q}}}\,\varphi = \varepsilon^{m}_{\dot{\alpha}}\overline{\mathcal{Q}}^{\dot{\alpha}}_{m}\varphi, \quad \boldsymbol{\delta}^{(2,0)}_{\bar{\varepsilon}\mathcal{Q}}\,\varphi = \bar{\varepsilon}_{m\dot{\alpha}}\,\mathcal{Q}^{m\dot{\alpha}}\,\varphi,$$

with

$$\overline{\mathcal{Q}}^{\dot{\alpha}}_{m} = \frac{1}{\partial^{+}}\begin{pmatrix}\bar{q}_m\bar{\partial}' + (q\boldsymbol{C})_m\partial \\ \bar{q}_m\bar{\partial} - (q\boldsymbol{C})_m\partial'\end{pmatrix}, \quad \mathcal{Q}^{m\dot{\alpha}} = \frac{1}{\partial^{+}}\begin{pmatrix} q^m\partial + (\bar{q}\boldsymbol{C})^m\bar{\partial}' \\ -q^m\partial' + (\bar{q}\boldsymbol{C})^m\bar{\partial}\end{pmatrix},$$

so that

$$\{\overline{\mathcal{Q}}^{\dot{\alpha}}_{m},\,\mathcal{Q}^{n\dot{\beta}}\} = -i\sqrt{2}\,\epsilon^{\dot{\alpha}\dot{\beta}}\delta^{n}_{m}\frac{\partial\bar{\partial} + \partial'\bar{\partial}'}{\partial^{+}}.$$

The transverse space rotations appear in right-hand side of the anticommutator,

$$\left\{\overline{\mathcal{Q}}^{\dot{\alpha}}_{m},\,Q^{n\alpha}\right\} = i\sqrt{2}\delta^{n}_{m}\nabla^{\dot{\alpha}\alpha}. \tag{15}$$

In superconformal theories, the dynamical supersymmetries generate by commutation all other dynamical transformations; for example the light-cone Hamiltonian, a derived quantity in all supersymmetric theories, is obtained from the anticommutator,

$$\left[\boldsymbol{\delta}^{(2,0)}_{\varepsilon\overline{\mathcal{Q}}},\,\boldsymbol{\delta}^{(2,0)}_{\bar{\varepsilon}'\mathcal{Q}}\right] = \boldsymbol{\delta}_{\varepsilon\bar{\varepsilon}'\mathcal{P}^{-}}. \tag{16}$$

3.3. *Self-dual string currents*

The interacting $(2,0)$ theory has defied all attempts at description, although many of its properties are well known [12].

The "free" chiral superfield encodes the Nambu–Goldstone modes of an eleven-dimensional supergravity solution living on one M5 brane. The interacting $(2,0)$ theory is the limit of a (non-conformal) theory with a number of separated M5 branes in mutual interaction through M2 branes. The B-field of each M5 brane is self-dual, and interacts with the closed string current of the M2 branes. In the limit where the M5 branes collapse on top of one another, the theory becomes conformal, and the $(2,0)$ theory emerges.

In the search for its formulation [13], several questions demand an answer. The compactification of $(2,0)$ on a torus yields the $\mathcal{N}=4$ SYM theory in $D=4$ and explains its S-duality. The SYM gauge indices cannot be carried by the B-field because its topological character forbids non-Abelian generalizations [14]. Where are the $\mathcal{N}=4$ gauge indices hiding in the $(2,0)$ theory?

In six dimensions, the B-field's canonical mass dimension is two, and its energy-momentum tensor is traceless, so that it is conformal (like the gauge potentials in four dimensions). The unique conformal coupling to closed string surface current is

$$\int d^6x \mathcal{J}_{\mu\nu}(x)\, B^{\mu\nu}(x), \tag{17}$$

where

$$\begin{aligned}\mathcal{J}^{\mu\nu}(x) &= \int d\tau d\sigma \Sigma^{\mu\nu} \delta^{(6)}(x_\mu - z_\mu(\sigma,\tau)),\\ &= \int d\tau d\sigma \left[\frac{\partial z^\mu}{\partial\tau}\frac{\partial z^\nu}{\partial\sigma} - \frac{\partial z^\mu}{\partial\sigma}\frac{\partial z^\nu}{\partial\tau}\right] \delta^{(6)}(x_\mu - z_\mu(\sigma,\tau)).\end{aligned}$$

The B-field, after disposing of the gauge-dependent flotsam, contains six transverse degrees of freedom, split by duality into two triplets of $SU(2)_L \times SU(2)_R$, as $(\mathbf{1},\mathbf{3})+(\mathbf{3},\mathbf{1})$.

We need to separate self-dual and anti-self-dual closed string surface currents. The transverse string coordinates, $z_i(\sigma,\tau)$ are no help because they transform symmetrically as $(\mathbf{2},\mathbf{2})$, but can be expressed as products of left- and right-handed spinors (twistors?).

Introduce $SU(2)_L$ spinors $\sim (\mathbf{2},\mathbf{1})$,

$$\boldsymbol{\lambda} = \begin{pmatrix}\lambda_1\\ \lambda_2\end{pmatrix} \to \mathcal{U}_L\boldsymbol{\lambda}, \quad \overline{\boldsymbol{\lambda}} = \begin{pmatrix}-\lambda_2^*\\ \lambda_1^*\end{pmatrix} \to \mathcal{U}_L\overline{\boldsymbol{\lambda}},$$

and $SU(2)_R$ spinors $\sim (\mathbf{1},\mathbf{2})$,

$$\boldsymbol{\rho} = \begin{pmatrix}\rho_1\\ \rho_2\end{pmatrix} \to \mathcal{U}_R\boldsymbol{\rho}, \quad \overline{\boldsymbol{\rho}} = \begin{pmatrix}-\rho_2^*\\ \rho_1^*\end{pmatrix} \to \mathcal{U}_R\overline{\boldsymbol{\rho}}.$$

The two matrix combinations,

$$\mathcal{X} \equiv \boldsymbol{\lambda}\boldsymbol{\rho}^\dagger + \bar{\boldsymbol{\lambda}}\bar{\boldsymbol{\rho}}^\dagger = \begin{pmatrix} \lambda_1\rho_1^* + \lambda_2^*\rho_2 & \lambda_1\rho_2^* - \lambda_2^*\rho_1 \\ \lambda_2\rho_1^* - \lambda_1^*\rho_2 & \lambda_2\rho_2^* + \lambda_1^*\rho_1 \end{pmatrix}$$

$$\mathcal{Y} \equiv \bar{\boldsymbol{\lambda}}\boldsymbol{\rho}^\dagger - \boldsymbol{\lambda}\bar{\boldsymbol{\rho}}^\dagger = \begin{pmatrix} -\lambda_2^*\rho_1^* + \lambda_1\rho_2 & -\lambda_2^*\rho_2^* - \lambda_1\rho_1 \\ \lambda_1^*\rho_1^* + \lambda_2\rho_2 & \lambda_1^*\rho_2^* - \lambda_2\rho_1 \end{pmatrix}$$

have the form of the transverse coordinate matrix X of Eq. (9), with complex conjugation defined as $(\rho_\alpha\lambda_{\dot\beta})^* = (\rho_\alpha^*\lambda_{\dot\beta}^*)$. These spinors could carry indices which are summed for in the coordinates. The surface current two-forms then split up,

$$(\acute{\mathcal{X}}^\dagger\dot{\mathcal{X}} - \dot{\mathcal{X}}^\dagger\acute{\mathcal{X}}) \to \mathcal{U}_R(\acute{\mathcal{X}}^\dagger\dot{\mathcal{X}} - \dot{\mathcal{X}}^\dagger\acute{\mathcal{X}})\mathcal{U}_R^\dagger,$$

$$(\acute{\mathcal{X}}\dot{\mathcal{X}}^\dagger - \dot{\mathcal{X}}\acute{\mathcal{X}}^\dagger) \to \mathcal{U}_L(\acute{\mathcal{X}}\dot{\mathcal{X}}^\dagger - \dot{\mathcal{X}}\acute{\mathcal{X}}^\dagger)\mathcal{U}_L^\dagger,$$

where "dot" and "prime" ($\dot{}$ and $\acute{}$) denote differentiation with respect to τ and σ, which transform as self-dual $((\mathbf{1},\mathbf{3}))$ and anti-self-dual $((\mathbf{3},\mathbf{1}))$ combinations, respectively. Both have the form,

$$\begin{pmatrix} iv & w \\ -\overline{w} & -iv \end{pmatrix}, \quad v = (\Sigma_{12} \mp \Sigma_{34}), \quad w = (\Sigma_{31} \mp \Sigma_{24} + i\Sigma_{32} \mp i\Sigma_{41})$$

which label the self-dual (anti-self-dual) triplet components. The self- and anti-self-dual currents are naturally separated by group theory. Assuming that $\boldsymbol{\rho}$, $\boldsymbol{\lambda}$ anticommute, we obtain the tensor structure of the $SU(2)_R$ vector current,

$$\begin{aligned} \vec{\mathcal{J}}_R \sim\; & \dot{\boldsymbol{\lambda}}^\dagger\acute{\boldsymbol{\lambda}}\,\boldsymbol{\rho}^\dagger\vec{\tau}\boldsymbol{\rho} + \dot{\boldsymbol{\lambda}}^\dagger\boldsymbol{\lambda}\acute{\boldsymbol{\rho}}^\dagger\vec{\tau}\boldsymbol{\rho} - \boldsymbol{\lambda}^\dagger\dot{\boldsymbol{\lambda}}\boldsymbol{\rho}^\dagger\vec{\tau}\acute{\boldsymbol{\rho}} + \dot{\boldsymbol{\lambda}}^\dagger\bar{\boldsymbol{\lambda}}\,\acute{\bar{\boldsymbol{\rho}}}^\dagger\vec{\tau}\boldsymbol{\rho} - \dot{\bar{\boldsymbol{\lambda}}}^\dagger\boldsymbol{\lambda}\,\boldsymbol{\rho}^\dagger\vec{\tau}\acute{\bar{\boldsymbol{\rho}}} \\ & - \boldsymbol{\lambda}^\dagger\boldsymbol{\lambda}\dot{\boldsymbol{\rho}}^\dagger\vec{\tau}\acute{\bar{\boldsymbol{\rho}}} - \frac{1}{2}\boldsymbol{\lambda}^\dagger\bar{\boldsymbol{\lambda}}\dot{\bar{\boldsymbol{\rho}}}^\dagger\vec{\tau}\acute{\boldsymbol{\rho}} - \frac{1}{2}\bar{\boldsymbol{\lambda}}^\dagger\boldsymbol{\lambda}\,\dot{\boldsymbol{\rho}}^\dagger\vec{\tau}\acute{\bar{\boldsymbol{\rho}}} - (\cdot \leftrightarrow \prime) \sim (\mathbf{1},\mathbf{3}). \end{aligned}$$

The $SU(2)_L$ triplet follows from $\boldsymbol{\lambda} \leftrightarrow \boldsymbol{\rho}$,

$$\begin{aligned} \vec{\mathcal{J}}_L \sim\; & \dot{\boldsymbol{\rho}}^\dagger\acute{\boldsymbol{\rho}}\,\boldsymbol{\lambda}^\dagger\vec{\tau}\boldsymbol{\lambda} + \dot{\boldsymbol{\rho}}^\dagger\boldsymbol{\rho}\acute{\boldsymbol{\lambda}}^\dagger\vec{\tau}\boldsymbol{\lambda} - \boldsymbol{\rho}^\dagger\dot{\boldsymbol{\rho}}\boldsymbol{\lambda}^\dagger\vec{\tau}\acute{\boldsymbol{\lambda}} + \dot{\boldsymbol{\rho}}^\dagger\bar{\boldsymbol{\rho}}\,\acute{\bar{\boldsymbol{\lambda}}}^\dagger\vec{\tau}\boldsymbol{\lambda} - \dot{\bar{\boldsymbol{\rho}}}^\dagger\boldsymbol{\rho}\,\boldsymbol{\lambda}^\dagger\vec{\tau}\acute{\bar{\boldsymbol{\lambda}}} \\ & - \boldsymbol{\rho}^\dagger\boldsymbol{\rho}\dot{\boldsymbol{\lambda}}^\dagger\vec{\tau}\acute{\bar{\boldsymbol{\lambda}}} - \frac{1}{2}\boldsymbol{\rho}^\dagger\bar{\boldsymbol{\rho}}\,\dot{\bar{\boldsymbol{\lambda}}}^\dagger\vec{\tau}\acute{\boldsymbol{\lambda}} - \frac{1}{2}\bar{\boldsymbol{\rho}}^\dagger\boldsymbol{\rho}\,\dot{\boldsymbol{\lambda}}^\dagger\vec{\tau}\acute{\bar{\boldsymbol{\lambda}}} - (\cdot \leftrightarrow \prime) \sim (\mathbf{3},\mathbf{1}). \end{aligned}$$

If the spinors satisfy a free string equation, they separate into left- and right-movers. With both $\boldsymbol{\rho}$ and $\boldsymbol{\lambda}$ left or right movers, the dual currents vanish due to the prime–dot antisymmetry, but if $\boldsymbol{\rho}$ is a right mover and $\boldsymbol{\lambda}$ is a left mover, or *vice versa*, the currents do not vanish.

With $\dot{\boldsymbol{\lambda}} = \acute{\boldsymbol{\lambda}}$ and $\dot{\boldsymbol{\rho}} = -\acute{\boldsymbol{\rho}}$, the $SU(2)_R$ triplet reduces to

$$\vec{\mathcal{J}}_R \sim 2\dot{\boldsymbol{\lambda}}^\dagger \boldsymbol{\lambda}\, \dot{\boldsymbol{\rho}}^\dagger \vec{\tau} \boldsymbol{\rho} - 2\boldsymbol{\lambda}^\dagger \dot{\boldsymbol{\lambda}} \boldsymbol{\rho}^\dagger \vec{\tau} \dot{\boldsymbol{\rho}} + 2\dot{\boldsymbol{\lambda}}^\dagger \overline{\boldsymbol{\lambda}}\, \dot{\overline{\boldsymbol{\rho}}}^\dagger \vec{\tau} \boldsymbol{\rho} - 2\dot{\overline{\boldsymbol{\lambda}}}^\dagger \boldsymbol{\lambda}\, \boldsymbol{\rho}^\dagger \vec{\tau} \overline{\boldsymbol{\rho}}.$$

3.3.1. *Kinematical supersymmetry*

In the free theory, the self-dual B-field is a component of the tensor multiplet superfield, so that the self-dual string current is itself part of another chiral constrained superfield, and if the coupling (17) exists in the interacting theory, the spinors are themselves components of Viking superfields,

$$\begin{aligned}\Lambda^{(A,r)}_\alpha(y,\tau,\sigma) = {} & \frac{1}{\partial^+}\boldsymbol{\lambda}^{(A,r)}_\alpha + \frac{i}{\sqrt{2}}\,\theta^{mn}\,\overline{\boldsymbol{\psi}}^{(A,r)}_{\alpha mn} \\ & + \frac{1}{12}\theta^{mnpq}\,\epsilon_{mnpq}\,\partial^+\,\overline{\boldsymbol{\lambda}}^{(A,r)}_\alpha + \frac{i}{\partial^+}\,\theta^m\,\overline{\mathbf{D}}^{(A,r)}_{\alpha m} \\ & + \frac{\sqrt{2}}{6}\,\theta^{mnp}\,\epsilon_{mnpq}\,\mathbf{D}^{q\,(A,r)}_\alpha,\end{aligned}$$

with all components functions of y, τ, σ, where A is a gauge index, and r stands for other indices (needed because the number of degrees of freedom in the theory increases as the cube of the number of M5 branes).

The $SU(2)$ inner automorphism allows for the "inside-out" constraints,

$$\overline{d}_m\,\overline{d}_n\,\Lambda^{A,r}_\alpha = \frac{1}{2}\,\epsilon_{mnpq}\,d^p\,d^q\,\overline{\Lambda}^{A,r}_\alpha.$$

$\boldsymbol{\rho}^{A,r}$ is the first component of a second superfield,

$$P^{(A,r)}_{\dot\alpha}(y,\tau,\sigma) = \frac{1}{\partial^+}\,\boldsymbol{\rho}^{(A,r)}_\alpha(y,\tau,\sigma) + \cdots.$$

In summary, this analysis suggests a new spinor framework where the kinematical requirements of the $OSp(6,2|4)$ symmetry are automatically satisfied.

Expressing the string coordinates in terms of spinors may provide clues to the variables which generate the non-linear realization of $OSp(6,2|4)$. Much work remains to be done to vindicate this approach.

Acknowledgements

Many thanks to Professors Lars Brink and Sung-Soo Kim, my early collaborators in this work for many discussions. I wish to thank the Aspen Center for Physics (partially supported by NSF grant PHY-106629), the Kavli Institute for Theoretical Physics, and the SLAC theory group, where part of this work was performed. This research is partially supported by the Department of Energy Grant No. DE-FG02-97ER41029.

References

[1] Y. Nambu, Relativistic wave equation for particles with internal structure and mass spectrum, *Supplement of the Progress of Theoretical Physics* **37** (1966) 368.
[2] E. Majorana, *Nuovo Cim.* **9** (1932) 335.
[3] R. Casalbuoni, *Proceedings of Science* **EMC2006** (2006) 004. arXiv:hep-th/0610252.
[4] R. Dolen, D. Horn, and C. Schmid, *Phys. Rev.* **166** (1968) 1768–1781.
[5] G. Veneziano, Construction of a crossing-symmetric, Regge behaved amplitude for linearly rising trajectories, *Nuovo Cim.* **A57** (1968) 190–197.
[6] Many of Nambu's published and unpublished works appear in *Broken Symmetry: Selected Papers of Y. Nambu*, T. Eguchi and K. Nishijima, editors, World Scientific, Singapore, 1995.
[7] M. S. Virasoro, Subsidiary conditions and ghosts in dual resonance models, *Phys. Rev.* **D1** (1970) 2933–2936.
[8] M. Kalb and P. Ramond, *Phys. Rev.* **D9** (1974) 2273–2284; E. Cremmer and J. Scherk, *Nucl. Phys.* **B72** (1974) 117.
[9] L. Brink, O. Lindgren and B. E. W. Nilsson, The ultraviolet finiteness of the $N = 4$ Yang–Mills theory?, *Phys. Lett.* **B123** (1983) 323.
[10] S. Ananth, L. Brink, S.-S. Kim and P. Ramond, Non-linear realization of $PSU(2,2|4)$ on the light-cone, *Nucl. Phys.* **B722** (2005) 166.
[11] W. Nahm, *Nucl. Phys.* **B135** (1978) 149.

[12] E. Witten, Some comments on string dynamics, in *Future Perspectives in String Theory*, Los Angeles, 1995; M. R. Douglas and G. W. Moore, D-branes, quivers, and ALE instantons, (1996) hep-th/9603167; J. Distler and A. Hanany, *Nucl. Phys.* **B490** (1997) 75; J. M. Maldacena, *Adv. Theor. Math. Phys.* **2** (1998) 231, and *Int. J. Theor. Phys.* **38** (1999) 1113; O. Aharony, A. Hanany and B. Kol, *JHEP* **9801** (1998) 002; C. Grojean and J. Mourad, *Class. Quant. Grav.* **15** (1998) 3397; P. Arvidsson, E. Flink and M. Henningson, *JHEP* **0405** (2004) 048; J. H. Schwarz, *JHEP* **0411** (2004) 078; A. Basu and J. A. Harvey, *Nucl. Phys.* **B713** (2005) 136.

[13] M. Douglas, KITP Slides, http://online.kitp.ucsb.edu/online/qft-c14/douglas/.

[14] D. Gaiotto, A. Kapustin, N. Seiberg and B. Willett, *JHEP* **1502** (2015) 172. arXiv:1412.5148.

Chapter 14

Nambu at Work*

Peter G. O. Freund

Enrico Fermi Institute and Department of Physics
University of Chicago, Chicago, IL 60637 USA

Yoichiro Nambu, whose life and seminal contributions to Physics we celebrate here, went in 1952 to the Institute for Advanced Study in Princeton. Shortly after his arrival there, J. Robert Oppenheimer, the Institute's director, put Yoichiro and the other new arrivals on notice that though Albert Einstein was a professor at the Institute, and therefore had an office there, nobody was to disturb the great man without first receiving special permission personally from Oppie. Most people would spend a year or two in the same building with Einstein and then spend a whole lifetime regretting not to have met him. Yoichiro decided that he will meet Einstein, no matter what Oppie says. He knew Bruria Kaufmann, Einstein's assistant at that time, and with her help got to visit the great physicist. Einstein was very friendly and visibly happy that finally one of the young people had bothered to visit him. Einstein asked Yoichiro what was going on in particle physics, and was rather skeptical about separate nucleon and meson fields for which he saw no deeper reason.

---

*Based on a talk delivered on 16 November, 2015 in Osaka at the *Nambu's Century: International Symposium on Yoichiro Nambu's Physics.*

It is well-known that Einstein never learned to drive. Yoichiro offered him a ride, ran to his car to open the door, and from the driver's seat snapped a picture of Einstein, its composition worthy of a major photographer.

Nambu offered Einstein a ride in his car and took this picture of him.

I brought this episode up at the very beginning of this talk, because there is a certain resemblance between the way the major works of Einstein and Nambu were conceived. In the construction of a relativistic theory of gravitation, Einstein used as a guide Mach's principle [1]. Actually this principle was not really embedded in general relativity, the end-product of Einstein's work. Though many people [2] have tried since, to this day Mach's principle is not incorporated in the theory, which can be understood without any reference to this beautiful but somewhat vague and philosophical principle.

When constructing with Moo-Young Han [3] the color gauge theory of strong interactions, Yoichiro, as he told me, was also guided by a philosophical principle, the quasi-Hegelian so-called "three stages" principle of Mitsuo Taketani [4]. The physics of mesons and baryons was for strong interactions the first *phenomenological* stage, somewhat like the 19th century spectroscopic results were for quantum theory. Similarly the Gell-Mann–Zweig quarks [5] represented the second *substantive* stage, like Bohr's atom, but the deeper *essentialist* stage, Taketani's third and final stage, the full grown theory like Heisenberg and Schrödinger's quantum mechanics, was to involve something unexpected in the form of a non-abelian color SU(3) gauge theory. Yoichiro did not sympathize with Taketani's politics, but, though it may not be well-known, he subscribed to Taketani's three-stages philosophy for physics theories, and was convinced that it was useful to him in the color gauge work, and as we shall see, in his string theory work as well.

The idea of a gauge theory as the basis for strong interactions was in the back of Yoichiro's mind for quite some time. Before his proposal of the non-abelian color gauge theory, Yoichiro had tried to gauge the Pauli–Pürsey–Gürsey symmetry [6], which mixes spinorial Fermi fields with their charge conjugates, and which had played a role in Heisenberg's much heralded but flawed unified theory [7]. I remember a number of conversations on this topic. That did not lead to any interesting results, but color gauge theory did.

Heisenberg's unified theory also influenced Yoichiro's discovery of spontaneous symmetry breaking. Besides its origin in Nambu's work on BCS [8] theory, the four-Fermi interaction starting point of the Nambu–Jona–Lasinio papers [9] reflects also the influence of Heisenberg. In fact, Heisenberg had thought of breaking the symmetries of his four-Fermi lagrangian and made an inspired analogy with ferromagnetism [7], but he missed the crucial feature of such a symmetry breaking mechanism: the appearance of a Nambu–Goldstone boson, in this case the so-called spin-wave or magnon, discovered by Heisenberg's former student Felix Bloch, and further studied in an important paper by Holstein and Primakoff [10], though the wide generality of this idea was not recognized by any of them.

It is amusing that the four-Fermi interaction in the BCS theory of superconductivity, with the electron as the fermion, is what gave that theory its "from basic principles" cachet. The bosonic Ginzburg–Landau (GL) theory [11] was viewed by the solid-state-physics community as nothing more than a phenomenological model, at least until Gor'kov showed [12] that GL can be derived from BCS.

In particle physics the situation is reversed. Shortly after Nambu's fundamental papers, in which the degeneracy of the vacuum and the appearance of massless bosons was discovered, Goldstone (giving full credit to Nambu) recast [13] their approach in an elegant bosonic classic field theory language, which became the preferred version of the particle physics community. Goldstone then proved his brilliant and far-reaching theorem [14], which governs the appearance and number of the massless particles. These are the reasons why these massless particles became known as "Goldstone bosons," and only in recent times were they given the more appropriate name of Nambu-Goldstone bosons by the particle physics community.

For decades, in Chicago, at the weekly Enrico Fermi Institute theory seminar, some young physicist freshly out of graduate school would give a talk about "Goldstone bosons" with Nambu sitting in the first row, and not batting an eye. The beauty of all this is that in the long run, the history of physics always sorts itself out in a fair way.

When gauge fields were added to the mix, this led to the beautiful Englert–Brout–Higgs [15] phenomenon, and here again the earlier Englert-Brout approach was phrased in a quantum field theoretic language, whereas Higgs' version used Goldstone's classic field theory language. Again, the particle physics community opted for this simpler version and named the outcome the Higgs phenomenon, and again only in recent times has the more appropriate name of Englert–Brout–Higgs phenomenon been more generally adopted.

In those days before the internet and even the Xerox copier, Englert and Brout mailed a copy of their preprint to Yoichiro, who liked it and came to my office to discuss it. We, of course, wanted to see whether this mechanism could be used to give mass to the by then well-known vector mesons, assuming that they were the gauge bosons of a Sakurai-type [16] SU(3)-flavor-gauge theory of strong interactions. By lunchtime, we had convinced ourselves that if this was the mechanism by which these vector mesons acquired their mass, then the K* had to be much heavier than the ρ, and only the charged ρ's were massive, while their neutral partner would stay massless. This was all seriously at odds with the experimental situation, and we concluded that, though it was a pity, this mechanism was of no use. It remained to be discovered by Weinberg [17] and Salam [18] that this mechanism was designed not for giving mass to some hadrons, but for the higher task of giving mass to the carrier bosons of the weak interactions.

Now let me move forward in time to the late sixties, the era of the birth of string theory. Again a phenomenological stage, the so-called two-component duality [19], started the game, the two components being destined to correspond to the open and closed strings. It was followed by the substantialistic stage of the Veneziano model [20], and in the hands of Nambu [21], Susskind [22] and Nielsen [23] it took on the *essentialistic* form of string theory.

I remember one morning in Yoichiro's office, Yoichiro telling me about his insight. We often tested our ideas on each other, and I must mention that quite a few of my ideas became much clearer after they were "tested" on Yoichiro. To appreciate Yoichiro's presentation of his new results one had to know him and be familiar with his personality. He would start by mentioning some very well-known results and putting on the blackboard in his beautiful handwriting a list of famous names. If you did not know Yoichiro, all this had the effect of making you think that he was going to tell you about a small detail in some well-known previous theory, a detail which maybe wasn't even new. But then, after about ten minutes or so, suddenly a new and often very deep approach would be coming to life in front of your eyes. It was easy to understand what he was saying, but where all this beautiful new stuff was coming from remained wrapped in total mystery.

As one brief example, when he arrived at string theory, what took him by quite some surprise was the exponential vertex operator, the normal-ordered version $\mathbf{:exp[ik\ X(0,\tau)]:}$. He came up with a very simple argument for this vertex: If I want to carry out a space translation by $\mathbf{\Delta x}$, I use the familiar operator $\mathbf{exp(i\Delta xP)}$, with $\mathbf{P}$ the momentum operator, the generator of space translations. But here I want to modify not the position by $\mathbf{\Delta x}$, but rather the momentum of the state by $\mathbf{k}$. So, by analogy, I use the operator $\mathbf{:exp[ik\ X(0,\tau)]:}$ with $\mathbf{X(0,\tau)}$ generating the required momentum "translation" during an interaction. Justifiably, he was proud of this argument. After all, this Nambu vertex turned out to be a mathematically very useful and richly endowed object.

There are two styles of doing seminal work in theoretical physics. In one style, an extremely clever physicist familiar with current theories and experimental results realizes that there is something missing for all this to gel and comes up with the theory that does the trick. This is the style of Einstein, of Heisenberg, of Yukawa and of Gell-Mann. Anybody can understand how they got to their theories. You can do both, admire and use the end result, *and* understand perfectly well where their ideas came from.

In the other style, the extremely clever physicist realizes that something is missing for a theory to attain its deeper meaning, and then supplies this missing element. This is the style of Einstein, of Dirac, of Feynman and of Nambu. You can again admire and use their end results, but how they ever got these ideas remains, as I said, wrapped in mystery. Yes, Einstein did work in both styles.

At this point it may be appropriate to say a few words about Nambu the man. Beside his marvelous modesty, Yoichiro was fully attuned to American ways, while still hanging on to his Japanese customs. In Japan politeness and conflict avoidance dictate a minimal use of the word "no." In Yoichiro's case this meant no use at all of this word. No matter how preposterous the request, after a pause he would always answer with a "yes," but the length of the pause measured his wish to say "no," the longer the pause, the more negative the response. Yoichiro's definition of the word "no" is a "yes" spoken after an infinitely long pause.

This led to some funny situations when Yoichiro became chairman of our Physics Department. It also led to one of Yoichiro's doctoral students staying around for many years and when the student did not leave on his own, Yoichiro let him graduate, simply because he could not bring himself to tell the student that he was *not* fit for doing theoretical physics.

I should mention here that Yoichiro had a number of excellent students: Lou Clavelli, Sumit Das, Savas Dimopoulos, Tony Gherghetta, Markus Luty, Burt Ovrut, Richard Prange, Jorge Willemsen, Motohiko Yoshimura, to name but a few.

Of the many brilliant scientists I met over the years, Yoichiro is one of the very few about whom I am certain was a genius.

Appendix

It may be somewhat unusual to find an appendix in an article of this kind, but I would like to reproduce here part of an email message I got from Yoichiro on 5 December, 2013. That message starts with some comments on Japanese politics, which I shall omit here, and then goes on to a science question:

> *"Dear Peter*
>
> ....
>
> *By the way I would like to ask you a question. In the course of studying Bode's law I found a mysterious paper on a derivation of the Schroedinger equation from classical dynamics by regarding time evolution not as a simple time derivative but as a stochastic process. Nelson E., Derivation of the Schrödinger equation from Newtonian mechanics, Physical Review, Vol. 150, No. 4, 1079–1084 (1966).*
> *I have not understood the paper yet. (There are careless errors in equations.) Did you know about it?*
> *YN"*

For completeness, here is my answer:

> *"Dear Yoichiro,*
>
> *Yes, I have seen Nelson's derivation of the Schroedinger equation. I always thought of it as a kind of perverse vindication of Hamilton's and Lagrange's work. What I mean to say is that there are three approaches to classical mechanics, Newton's, Hamilton's and Lagrange's. The latter two, though originally developed as pure mathematical physics, contained the remarkable optics-mechanics analogy, and naturally led to the Heisenberg, the Schroedinger and the Feynman approaches to QM. It would be hard to see how*

QM would have been discovered, had Hamilton and Lagrange not done their work in the nineteenth century. But that leaves open the question as to whether QM could have been arrived at, had Newton's approach been the only one known. Nelson answered that question in the affirmative, but I doubt that QM would have been discovered as "naturally" and as early, had the Newton/Nelson approach been the only one available. I find Nelson's work to be more of the "for completeness' sake" type, although I admit that it may still provide useful clues in the future.
Best
Peter"

Bode's law — or more accurately the Titius–Bode law — mentioned here, states that in our solar system, in astronomical units, the semi-major axes of the planets' orbits are given by the formula

$$a(n) = 0.4 + 0.3 \cdot 2^n \quad \text{with } n = -\infty, 0, 1, 2, 3, 4, \ldots.$$

so that $n = 1$, $a(1) = 1$ corresponds to the Earth, as befits astronomical units. This law was arrived at in the eighteenth century and it is obeyed to within an error of 5% by the first eight planets. For Neptune the law's prediction is too large by some 30%, and for Pluto it is too large by almost a factor 2, but improved versions of the law that accommodate Neptune and Pluto have been proposed, which are valid for other planetary or satellite systems as well. The relevance of a Schrödinger-type equation for such systems has been observed by Albeverio, Blanchard, Høegh-Krohn [24] and others. These issues are reviewed and expanded in a paper by Scardigli [25]. In the simplest way, the appearance of a Schrödinger-type equation can be seen as follows [25]. In a Bohr-like model let us require the usual equality of the gravitational attraction and centrifugal forces,

$$GMm/r^2 = mv^2/r \tag{1}$$

and then impose the new type of "quantization" condition

$$J/m = vr = se^{\lambda n} \tag{2}$$

instead of the Bohr quantization

$$J = \hbar n. \tag{3}$$

Here M is the sun's mass, m is the mass of the planet, which on account of the quantization condition (2) and of the equivalence principle which guarantees the appearance of the same mass m on both sides of Eq. (1), ultimately cancels out, v is the planet's velocity, r its distance from the sun, G is Newton's constant, J is the orbital angular momentum whereas s and λ are new parameters. Equations (1) and (2) yield a law with n in the exponent,

$$r_n = r_0 e^{2\lambda n} \quad \text{with } r_0 = s^2/GM, \quad n = 1, 2, 3, \ldots. \tag{4}$$

just like the improved versions of the Titius–Bode law. Comparing with experimental data on various planetary or satellite systems, the parameter s, unlike Planck's constant which it replaces, does not take a universal value, but changes from system to system. However the parameter λ remains essentially the same [25].

Finally, just like Bohr's atom, requires the well-known Schrödinger equation to make sense, so this Titius-Bode type system, with its different angular momentum operator, yields a different Schrödinger-like equation [24, 25]. There are of course no quantum jumps from one level to another, no superposition principle, etc.... The stochastic feature of this Schrödinger-type equation originates in the classical chaotic dissipative processes which, over a sufficiently long time, stabilize the original dust in the proto-planetary system in Titius-Bode orbits. At a deeper level these ideas are connected [25] to the "determinism beneath quantum mechanics" advocated by 't Hooft [26]. Given these facts, I wonder where Yoichiro was headed when pursuing these ideas. As always, we are probably going to find out ten years from now why he was on the right path already in 2013.

References

[1] E. Mach, *Die Mechanik in Ihrer Entwicklung Historisch-Kritisch Dargestellt*, Brockhaus, Leipzig, 1912.
[2] M.Y. Han and Y. Nambu, Three-Triplet Model with Double SU(3) Symmetry, *Phys. Rev.* **139** B1006 (1965).
[3] F. Gürsey, Reformulation of General Relativity in Accordance with Mach's Principle, *Ann. Phys. NY*, **24**, 211 (1963).

[4] M. Taketani and M. Nagasaki, *The Formation and Logic of Quantum Mechanics*, Vols. I-III, World Scientific, Singapore, 2001.
[5] M. Gell-Mann, A Schematic Model of Baryons and Mesons, *Phys. Lett.* **8**, 151 (1964); G. Zweig, CERN Preprint TH401 (1964).
[6] W. Pauli, On the Conservation of Lepton Charge, *Nuovo Cim.* **6**, 204 (1957); D.L. Pürsey, Invariance Properties of Fermi Interactions, *Nuovo Cim.* **6**, 266 (1957); F. Gürsey, Relation of Charge Independence and Baryon Conservation to Pauli's Transformation, *Nuovo Cim.* **7**, 417 (1958).
[7] H.-P. Dürr, W. Heisenberg, H. Mitter, S. Schlieder, and K. Yamazaki, Zur Theorie der Elementarteilchen, *Z. Naturforschung* **14**, 441 (1959).
[8] J. Bardeen, L.N. Cooper and J.R. Schrieffer, Theory of Superconductivity, *Phys. Rev.* **108**, 1175 (1957); Y, Nambu, Quasi-Particles and Gauge Invariance in the Theory of Superconductivity, *Phys. Rev.* **117**, 648 (1960).
[9] Y. Nambu and G. Jona-Lasinio, Dynamical Model of Elementary Particles Based on an Analogy with Superconductivity, I and II, *Phys. Rev.* **122**, 345 (1961); ibid. **124**, 246 (1961).
[10] F. Bloch, Zur Theorie des Ferromagnetismus, *Z. f. Physik* **61**, 206 (1930); T. Holstein and H. Primakoff, Field Dependence of the Intrinsic Domain Magnetization of a Ferromagnet, *Phys. Rev.* **58**, 1098 (1940).
[11] V.L. Ginzburg and L.D. Landau, On the Theory of Superconductivity, *Zh. Eksp. Teor. Fiz.* **20**, 1064 (1950).
[12] L.P. Gor'kov, *Microscopic Derivation of the Ginzburg-Landau Equations in the Theory of Superconductivity*, Sov. Phys. JETP, **36(9),** 1364 (1959).
[13] J. Goldstone, Field Theories with Superconductor Solutions, *Nuovo Cim.* **19** 154 (1961).
[14] J. Goldstone, A. Salam and S. Weinberg, Broken Symmetries, *Phys. Rev.* **127**, 965 (1962).
[15] F. Englert and R. Brout, Broken Symmetry and the Mass of Gauge Vector Mesons, *Phys. Rev. Lett.* **13**, 321 (1964); P.W. Higgs, Broken Symmetries and the Masses of Gauge Bosons, *Phys. Rev. Lett.* **13**, 508 (1964).
[16] J.J. Sakurai, Theory of Strong Interactions, *Ann. Phys. NY* **11**, 1 (1960).
[17] S. Weinberg, A Model of Leptons, *Phys. Rev. Lett.* **19**, 1264 (1967).
[18] A. Salam, Weak and Electromagnetic Interactions, *Proc. 8th Nobel Symposium*, ed. N. Svartholm, p. 367, Wiley NY, 1968.
[19] R. Dolen, D. Horn and C. Schmid, Prediction of Regge Parameters of ρ Poles from Low-Energy πN Data, *Phys. Rev. Lett.* **19**, 402 (1967); P.G.O. Freund, Finite Energy Sum Rules and Bootstraps, *Phys. Rev.*

Lett. **20**, 235; H. Harari, Pomeranchuk trajectory and its Relation to Low-Energy Scattering Amplitudes, *Phys. Rev. Lett.* **20**, 1395 (1968).

[20] G. Veneziano, Construction of a Crossing-Symmetric, Regge-Behaved Amplitude for Linearly Rising Trajectories, *Nuovo Cim.* **A57**, 190 (1968).

[21] Y. Nambu, Quark Model and the Factorization of the Veneziano Amplitude, in *Symmetries and Quark Models*, ed. R. Chand, p. 269, Gordon and Breach NY, 1970; Lectures at Copenhagen Symposium, 1970.

[22] L. Susskind, Dual Symmetric Theory of Hadrons, *Nuovo Cim.* **69A**, 457 (1970).

[23] H. B. Nielsen, *An Almost Physical Interpretation of the Dual N-Point Function*, Nordita Report, unpublished 1969.

[24] S. Albeverio, Ph. Blanchard and R. Høegh-Krohn, A Stochastic Model for the Orbits of Planets and Satellites: An Interpretation of the Titius-Bode Law, *Expo. Math.* **4**, 365 (1983).

[25] F. Scardigli, *A Quantum-Like Description of the Planetary Systems* [arXiv: gr-qc/0507046].

[26] G. 't Hooft, *Determinism Beneath Quantum Mechanics*, [arXiv: quant-ph/0212095].

Chapter 15

Yoichiro Nambu

H. B. Nielsen

The Niels Bohr Institute, Copenhagen, Denmark
hbech@nbi.dk

Sitting down and going through my old folders in order to put them into boxes for sending to the Rigsarkivet (the archive for protecting old materials by the government of Denmark) I came to see a folder marked "Han–Nambu 2007" reminding me about some work Froggatt, Takanishi and I [1] did, which was inspired to such a degree by an idea of Moo-Young Han and Yoichiro Nambu, that I had put it in a folder. As we now know the Han–Nambu charge assignment [2,3] for the quarks turned out not to be the correct one, but it was so inspiring that I later wanted to use it for some speculations about why Nature should have chosen the Standard Model. Just the charge assignment part which we wanted to use for something else turned out not to be realized, but the great thing that Han and Nambu — and in a somewhat different way from them O. W. Greenberg [4] — proposed was what is now the color of the quarks and this latter is one real great discovery. Nambu has made so many things, which I would normally think I could not know them all. And I think others might have picked up the ideas on second hand. Are these ideas not often such that they could be spread in a few words by second hand, those ideas of Nambu's? At least you might think so after having learned them.

Works by Yoichiro Nambu include: the bound-state equation for relativistic quantum field theory in 1950, the year before the Bethe–Salpeter equation; prediction of the ω-meson in 1957; quantum field theory for superconductivity in 1960; principle of spontaneous symmetry breaking and the mass of elementary particles in 1960 and 1961; introduction of the color degree of freedom in 1965; non-Abelian gauge theory for strong interaction in 1965; introduction of the string theory in 1970; and the generalized Hamilton dynamics in 1973.

1. The string discovery

Nambu was one of us [5–9] who independently discovered that the Veneziano model(s) were indeed amplitudes for scattering of strings. For early development of string theory see also [10] celebrating the birth of string theory.

In his talk [5] at the International Conference on Symmetries and Quark Models titled "Quark model and the factorization of the Veneziano amplitude", he calculated the factorization into resonances of a Veneziano amplitude for an arbitrary number of external spin zero particles. As one may see from the question of Breit after the talk "Why do you have factorization?" it may not have been so obvious at that time, that one should have factorization. He introduced the state vector description of resonances in his equation (7)

$$\sum \langle n| \ldots |l\rangle\langle l|\Gamma|m\rangle\langle m|\Gamma|k\rangle\langle k|1\rangle \tag{1}$$

as the wished-for factorization, when a chain of resonances occur simultaneously for $\alpha(s_{12}) = k$, $\alpha(s_{123}) = m$, $\alpha(s_{1234}) = l$, etc. Then these states — possibly a set of states that could contain a superfluous number of states — were evaluated to be described by an infinite series of harmonic oscillators enumerated by the natural numbers for each dimension. Here the dimensions meant at first both space and time dimensions, and Nambu even needed an extra one, so in the experimental case of dimensions he needed 5. He wrote by analogy to ordinary field theory as a set of plane waves of fields ϕ_α. Then this suggested, he wrote, that the internal energy of a meson was analogous to that of a quantized string of a finite length (or a cavity resonantor for that matter) whose displacements were described by the field ϕ_α and found the free wave equation

$$(\Box - N - c)\Psi(x; \phi) = 0 \tag{2}$$

(where N is the Hamiltonian for the string).

Here the string is (almost) derived from the Veneziano model via the factorization.

This is somewhat different from my own approach, which (formally at least) circumvents the factorization and rather starts from an assumption of the string formulated as a chain of many constituents interacting with their neighbors in a quantum field theory, for these constituents are supposed to lead to "fish net diagrams", which in turn are shown to lead essentially to the Veneziano model. (My memory is that I rather needed the *two-dimensionality* of the important diagrams for the constituent interaction as a little extra assumption to get the shape/behavior of the integrand in the integral expression for the Veneziano model(s).)

You might say that Nambu [5, 6] and also Susskind [8] extracted the string more fully from the string model factorization while I [7] guessed the string and checked it without having to know about the factorization for my treatment. Presumably I only knew about factorization of the four point function, when I came to the string. But, well, actually in the notes prepared for the Copenhagen High Energy Symposium in August 1970, "Duality and hadrodynamics" [6], the last part of the article described a string picture much more like my own chain of constitutents. He talked about a number of constituents that was not only large but undetermined. This was connected to the Harari–Rosner schematical interpretation of duality [11, 12] and the figure illustrating the splitting of a meson into two in such a quark diagram was already presented in the first article [5]. He ended up, it seemed, describing the string as series of quark antiquark quark antiquark...

Indeed Nambu provided an explanation of the one-dimensionality very close to the modern thinking as chains of gluons in large number of color limit. But of course for good reasons there could be no gluons at that time. But the quarks — Han–Nambu quarks, what mattered was that they had two indices — interacted (really Nambu kept avoiding such a picture, and kept more to what he knew) with the neighbors in a chain making up the string.

1.1. *The Nambu–Goto action*

In the same Copenhagen High Energy Symposium contribution not presented [6] Nambu put forward what is now known as the Nambu–Goto action in the form

$$I = \int |d\sigma_{\mu\nu} d\sigma^{\mu\nu}|^{1/2} = \iint |2 \det g|^{1/2} d^2\xi \tag{3}$$

where

$$d\sigma^{\mu\nu} = G^{\mu\nu} d^2\xi, \tag{4}$$

$$G^{\mu\nu} = \frac{\partial(y^\mu, y^\nu)}{\partial(\xi^0, \xi^1)}, \tag{5}$$

whereas the line element is

$$ds^2 = g_{\alpha\beta} d\xi^\alpha d\xi^\beta, \tag{6}$$

$$g_{\alpha\beta} = (\partial y_\mu / \partial \xi^\alpha)(\partial y^\mu / \partial \xi^\beta). \tag{7}$$

Using this area action he already in this second paper [6] on the string showed the connection to the L_n operators being the gauge generating operations.

This was really a great progress in the understanding of the string in itself, which for instance I would only learn afterwords by directly or indirectly reading Nambu.

1.2. *Some other string work*

Already in the contribution to the Copenhagen High Energy Symposium [6] he mentioned the connection to the Hagedorn temperature spectrum and in fact made use of it, and considered a grand canonical ensemble of resonances. He managed to argue that the pomeron gets half the slope of that of the ordinary Regge trajectories.

He discussed the problems and especially that the formfactor for an infinite string becomes infinitely sharp in momentum space.

1.3. *String works with vortices*

Interesting for me was that Nambu took up the idea of Poul Olesens and myself [13] about identifying the strings with ours or rather Abrikosov's vortices [14]. Indeed he took the quarks to be magnetic monopoles and thus managed to get ends for these vortices [15–18], something Poul and I had not done ourselves.

2. Memory of Nambu

One of my memories about Yoichiro Nambu was the time he should have talked about the string in Copenhagen [6] but had the bad luck of being stranded in the desert and thus did not reach our SINBI conference in Copehagen. But he had written the text of the talk already and so it was accessible.

I feel I know Professor Nambu very well — presumably helped by knowing in some way or the other many of his work, and by having heard some stories in connection with my main advicer Ziro Koba being his near costudent at some time —, but honestly I only met him rather shortly each time, apart from the first time in Chicago in 1969 when we spent an afternoon dicussing. I have such great memories of that. I also remember an occasion later when he invited me very kindly in Paris to join him for a meal. That was a very memorable meal.

Yoichiro Nambu was an extremely kind person, a true genius who mostly showed it in writing.

References

[1] C. D. Froggatt, H. B. Nielsen and Y. Takanishi, Neutrino oscillations in extended anti-GUT model, *AIP Conf. Proc.* **540** (2000) 35, doi:10.1063/1.1328879 [hep-ph/0011168].

[2] Y. Nambu and M. Y. Han, Three triplets, paraquarks, and colored quarks, *Phys. Rev. D* **10** (1974) 674–683, doi: 10.1103/PhysRevD.10.674.

[3] M. Y. Han and Y. Nambu, Three triplet model with double SU(3) symmetry, *Phys. Rev.* **139** (1965) B1006–1010, doi: 10.1103/PhysRev.139.B1006.

[4] O. W. Greenberg, Spin and unitary spin independence in a paraquark model of baryons and mesons, *Phys. Rev. Lett.* **13** (1964) 598.

[5] Y. Nambu, "Quark model and the factorization of the Veneziano model amplitude" in *Proceedings of the International Conference on Symmetries and Quark Models*, Wayne State University, June 18–20, 1969, R. Chand (ed.) (Gordon and Breach, New York, 1970), pp. 269–277; reprinted in *Broken Symmetry: Selected Papers of Y. Nambu*, T. Eguchi and K. Nishijima (eds.) (World Scientific, Singapore, 1995), pp. 258–277.

[6] Y. Nambu, Duality and hadrodynamics, lecture notes prepared for Copenhagen summer school, 1970; repoduced in *Broken Symmetry: Selected Papers of Y. Nambu*, T. Eguchi and K. Nishijima (eds.) (World Scientific, Singapoore 1995), p. 280.

[7] H. B. Nielsen, A physical interpretation of the integrand of the n-point Veneziano model (1969) (Nordita preprint); An almost physical interpretation of the integrand of the n-point Veneziano model, preprint at Niels Bohr Institute; a paper presented at the 15th International Conference on High Energy Physics, Kiev 1970, p.445.

[8] L. Susskind, Structure of hadrons implied by duality, *Phys. Rev. D* **1** (1970) 1182–1186.

[9] L. Susskind, Dual symmetric theory of hadrons, *Nuovo Cimento A* **69** (1970) 457–496.

[10] A. Cappeli, E. Castellani, F. Colomo and P. Di Vecchia (eds.), *The Birth of String Theory* (Cambridge University Press, 2012).

[11] J. L. Rosner, Graphical form of duality, *Phys. Rev. Lett.* **22** (1969) 689.

[12] H. Harari, Duality diagrams, *Phys. Rev. Lett.* **22** (1969) 562.

[13] H. B. Nielsen and P. Olesen, Vortex line models for dual strings, *Nucl. Phys. B* **61** (1973) 45.

[14] A. A. Abrikosov, On the magnetic properties of superconductors of the second group, *Sov. Phys. JETP* **5** (1957) 1174.

[15] Y. Nambu, String-like configurations in the Weinberg–Salam theory, *Nucl. Phys. B***130** (1977) 505, doi: 10.1016/0550-3213(77)90252-8.

[16] Y. Nambu, Magnetic and electric confinement of quarks, *Phys. Rept.* **23** (1976) 250–253, doi: 10.1016/0370-1573(76)90044-2.

[17] Y. Nambu, Strings, monopoles and gauge fields, *Phys. Rev. D* **10** (1974) 4262, doi: 10.1103/PhysRevD.10.4262.

[18] Y. Nambu, Effective abelian gauge fields, *Phys. Lett. B* **102** (1981) 149, doi: 10.1016/0370-2693(81)91051-0.

Appendix

Reminiscences of the Youthful Years of Particle Physics[1]

Yoichiro Nambu

University of Chicago

1. I graduated from the Physics Department of the University of Tokyo in 1943. So I am one of the so-called "wartime guys". The Pacific War started in my second year. The third year was cut short and I was drafted into the Army, ended up being assigned to a radar laboratory near a city called Takarazuka, where I was able to gain experience in practical problems of physics. As the war ended, I was fortunate to be able to return to the University of Tokyo as a postdoctoral associate joining a group of repatriates and younger students.

Among the latter were Z. Koba, Y. Miyamoto and T. Kinoshita. Under the guidance of Prof. Tomonaga of the nearby Tokyo Bunrika University and the RIKEN Laboratory, they were collaborating with him to develop his so-called super-many-time theory. But progress was slow because of the difficult living condition. As for me, I was living in my office with a few others, and fortunately across my desk was Koba. Watching his work I was gradually able to learn the Tomonaga theory.

In 1947, the big news of the Lamb shift and of the discovery of the pion reached Japan. Immediately we got involved in a fierce competition with American physicists to develop quantum electrodynamics, or QED. So I also formally joined the Tomonaga group and listened to his lectures. The essence of the Tomonaga theory is relativistic invariance and renormalization. Tomonaga called renormalization the principle of "hohki", which means either "giving up" or "broom". I did not know which he meant, and did not like the word any way.

In time I became able to pursue my own research. In 1950, under the recommendation of Tomonaga, I joined S. Hayakawa, Y. Yamaguchi, K. Nishijima of University of Tokyo and T. Nakano of Osaka University to form a theory group at the newly created science department of Osaka City University.

First I would like to talk about the influence of the so-called Sakata–Taketani philosophy in the immediate postwar period. They were collaborators with Yukawa in creating the meson theory but also developed and advocated a unique methodology. Taketani often came to see his friend S. Nakamura near our office, and expounded his theory before us. I need not explain here his "three stages theory", but we youngsters were overwhelmed by his persuasive eloquence. He also spoke against our preoccupation with theoretical ideas, emphasized to pay attention to experimental physics. It was a good advice against the tendency resulting from the success of the Yukawa theory as well as the miserable postwar condition which made it difficult to do experimental physics. I believe that this advice has come to make a big influence on my attitude toward physics.

Still now I look back with nostalgia toward the three years at Osaka City University. It was the period when in America the technical and intellectual energies accumulated during the war got explosively released toward peaceful enterprises. In contrast, we in Japan were not yet able to foresee the future, feeling hungry every day in a barack office of the department. Nevertheless, I had never felt and enjoyed so much the sense of freedom. Being a new department, there were no senior professors to defer to, no need for formal lectures as there were only two students. We were completely free to engage in research by ourselves. Gradually news started to come in from abroad of discoveries found in accelerators and cosmic rays. As the international communication became easier, so did the internal one. Under these stimulations we started to make our own original work, which we published in our own tabloid magazine together with reports from people who were able to go overseas.

One of the successes of our collaborative efforts at Osaka City University was the pair production theory of the newly discovered "strange particles". It was surprise to us that we novices could do this ahead of the physicists of the world. Not only our group but those in Tokyo and Kyoto have also started to make original contributions. But we did not last. In a few years our very success made us disperse to more established institutions. Thus in 1952, I was invited to the Institute for Advanced Study (IAS) in Princeton together with Toichiro Kinoshita on the recommendation of Tomonaga.

2. Contrary to my expectation, however, my stay at IAS turned out to be a mixture of heaven and hell. Among the physicists, there were J. R. Oppenheimer the director, A. Einstein, and various kinds of members: W. Pauli, A. Pais, F. Dyson, C. N. Yang, T. D. Lee, G. C. Wick, G. Kallen, W. Thirring, etc. The short term members like me were housed in barracks on campus so we became close to each other right away. It was a heavenly environment compared to the miserable life we left behind. On the other hand, I could not help but feel of fierce competition among us. Moreover, my plan to pursue the saturation properties of nuclear forces and the origin of the spin-orbit force hit a wall and stalled. It made me recall the poem by Takuboku Ishikawa:

> "*On a day when all my friends look better than I,*
> *I would bring home flowers to enjoy with my wife*"

At the end of the first year, I was allowed to stay for another year, but did not receive support during the recess period from spring to summer. Oppenheimer kindly offered to use the fund left by Yukawa, arranged for Kinoshita and me to be able to spend the period at the California Institute of Technology. I decided to bring my wife and son from Japan and meet them there. So the Kinoshita couple and I started to drive two cars together and cross the country to the South. On the way we could witness the miserable status of the blacks. This was the time right after the famous ruling of the

Supreme Court. We were also expecting some unpleasant incidents on the way, but did not experience them until we reached Pasadena and looked for apartments. (A similar incident also occurred later in Chicago.)

At CalTech there was a 500 GeV synchrocyclotron. I asked to join the experimental team but was politely declined. So I contented myself doing analysis of the $\gamma - \pi$ reaction. The famous Oppenheimer incident occurred in my second year at IAS. But I was not yet sufficiently informed of American politics to pay much attention to it. At the end of the two years, I wanted to stay a bit longer at a leading university in order to accomplish something. There was an offer but it was too distant. I searched for a research associate position but it was very scarce in those days. I almost gave up when an offer came from M. L. Goldberger of Chicago. I had already met him, and had also visited the university. But in those days the impression of the city was awful. During the war, America had also spent all its energies toward winning it, had no room to pay attention to domestic matters, I suppose. The commuter train from Hyde Park, the site of the university, to downtown was like that in Tokyo. I would never come here again, that was my impression. It is an irony of fate that we have come to live there. As our car approached Chicago, we had to pass the famous area of steel industry called Gary, Indiana. We felt as if driving into hell.

When I settled down at the Institute for Nuclear Study (INS) of the university, however, I began to feel that its atmosphere was like a paradise. The name has since changed, but it was formed together with its sister, the Institute for the Study of Metals (ISM, a wartime code name) in order to house people engaged in the Manhattan Project, and occupied two halves of a large building. I got to know all the people there right away, and was treated like a member of a big family. Every week Enrico Fermi held a discussion session with the theorists. I attended it a couple of times but soon he disappeared and passed away. The rest of INS members, besides Goldberger, included G. Wentzel, S. Allison, J. and M. Mayer, J. L. Marshall, H. Anderson, J. Simpson, H. Urey and S. Chandrasekhar. In the separate Physics Department were W. Zachariasen, M. Schein, and

R. Mulliken. There was also a diverse group from Japan. Tadao Fujii had just got a PhD in nuclear physics,. H. Miyazawa was a research associate with Wentzel, R. Kubo and R. Kikuchi, my classmate at Tokyo, were members of ISM. M. Koshiba was a research associate in Schein's group. Y. Nakagawa was helping Chandrasekhar, doing experiments on magnetohydrodynamics. These people were all living within walking distance from the university. All the time we got invited to homes of senior professors. Especially famous was the grand year-end party at the Mayers, where an old barn house was filled up with people, virtually with standing room only. Regrettably this custom is gone now, presumably because it has become common for both husbands and wives to have a job. Following the tradition created by the universal physicist Fermi, we had a weekly institute-wide seminar. It was called Quaker meeting because there were no fixed agenda, everyone was encouraged to stand up and discuss any ideas one has at the moment, even if only half-baked. The topics ranged from all areas of the institute, from nuclear physics, particle physics, cosmic rays, solar phenomena to the physics and chemistry of the universe. This was a great stimulation for me.

INS had a 450 MeV synchrocyclotron built by Fermi and Anderson. It is here that the first hadron resonance called δ was discovered, and its interpretation was first given by K. Brueckner and H. Miyazawa ahead of Fermi. Also it is well known that the neutron–electron interaction was measured here. Frequently I would climb down a ladder to the control room in the pit, and chat with people there. I became especially close to Anderson.

On the theory side, Chicago was then a mecca of dispersion theory. It originated from the work of Gell-Mann and Goldberger, which was just the right tool for analyzing the experimental data coming in. Under Goldberger were H. Miyazawa, R. Oehme and myself. The mathematical structure of the dispersion theory fascinated me, and enabled me to forget the two years of bad dream. Besides collaborating with Golberger's friends F. Low and G. Chew at the University of Illinois, I made interesting discoveries about the representation of the Green's function, etc. These were just mathematical problems, but I was elated when L. Landau quoted my work. At the Kiev conference in 1959 he greeted me there as well as at his institute in Moscow.

Nevertheless, I had not forgotten the Sakata–Taketani philosophy. I was clearly conscious of it when I predicted the existence of a neutral vector meson in order to account for the property of the nucleon form factors. But the reaction I received was very cool. Resistance against new particles was still strong. The pion was the sole carrier of the strong interaction, the δ was only a π-nucleon resonance, they thought. Although I assumed the δ to be a new elementary particle, I postulated a low mass for it in order to make it stable against strong forces. But I was not clearly distinguishing between resonances and elementary particles.

According to Taketani's theory, the 1950s and 1960s were the time of transition from phenomenology to model building. It was generally recognized that Nature had, besides gravity, three forces, namely electromagnetic, strong, and weak, while the particles consisted of leptons (e and μ) and hadrons.[a] Regarding the weak forces, the Fermi theory modified with $V - A$ and Cabibbo mixing was satisfactory at least for the baryons. On the other hand, we were still in the dark about the true nature of the strong forces. Of course the QED was a complete success for electromagnetism. The concept of renormalization has introduced a new paradigm for quantum field theory. It was clear, though, that this was not enough for understanding the strong forces.

3. At this point I will stop for a moment to discuss the problems of methodology. There were two competing groups which I will tentatively call reductionists and generalists. The reductionists first try to find a symmetry or regularity in the phenomena, and trace it to the properties of the fundamental particles and interactions, Gell-Mann being their spokesman. The SU(2), SU(3) and SU(6) symmetries of flavor were established this way. But they were only approximate symmetries. Exact symmetries like the spacetime symmetry and the gauge invariance of electromagnetism were established by the first

[a]The name *lepton* is due to L. Rosenfeld; *hadron* is due to L. D. Okun.

half of the 20th century. On the other hand, approximate internal symmetries have since made their appearance in particle physics. They may be their inherent character, but may also reflect people's mode of thinking that is sensitive to similarities and analogies. In this respect I recall the following remark by A. Salam.[b]

"Classical physical theories are profound. Take the second law of thermodynamics, for instance. Heat cannot flow spontaneously from a colder body to a hotter body. Compare this to what you have been doing. You propose some symmetry, and ten seconds later you are already trying to figure out how to break it."

The gauge principle was generalized by Yang and Mills, and has come to be believed as an ideal theory for all fundamental forces. But it was not clear how it could be applied to the short range forces of the actual world having only approximate symmetries. J. J. Sakurai made a brave but tentative attempt to challenge the question with his vector dominance theory. He came to Chicago right after his PhD from Cornell University. His energetic efforts to develop his theory led to some successes, but what was lacking in completing his program was the recognition that the fundamental strong force were between hypothetical quarks, and the symmetries could break spontaneously.

The generalists, on the other hand, tried to clarify the general properties of quantum field theory, and use them in analyzing individual problems, in view of the fact that the fundamental laws are not completely known yet. It led to various theoretical tools such as axiomatics, S-matrix, dispersion relations, and Regge poles. The S-matrix was originally introduced in the 1930s by Heisenberg in connection with nuclear problems. Among the generalists, a radical group represented by G. Chew went beyond S-matrix and advocated the principle of particle democracy and bootstrap. This meant that there is no fundamental Lagrangian, all hadrons are bound states of each other. His charismatic character attracted many followers.

From the present point of view, it is clear that the royal road of physics was that taken by the reductionists. Yukawa's meson did

---

[b]This is quoted by J. J. Sakurai in *Ann. Phys.* **11** (1960), p. 1.

what the S-matrix was not able to do. Thus the success of Yukawa's approach continued and eventually led to the present Standard Model. The road taken by the generalists, however, was not an idle one. Not only have the concepts like the CPT theorem been very useful for phenomenological analyses, but also today's superstring theory may be called its descendant. Thus the competition is still on, or rather they are moving toward unification, one might say.

Going back to the 1960s, I would first like to describe the social and political atmosphere of the time as I remember it. The 1960s were indeed an eventful decade. The assassination of the Kennedys was followed by the Vietnam war and the anti-war movement went hand-in-hand and gradually escalated. The anti-discrimination movement also followed the same path, leading to the assassination of Martin Luther King and the ensuing riots. At the same time, a rocket carrying man reached the moon. Hanging over all these events was the thickening dark cloud of the cold war and the atom bomb. The physicists had lost their purity because of their special role played in the War. This resulted in the big reward they received from the government. As their research budget reached astronomical levels, the accelerator energy kept going up following the Livingstone curve. but they also had to collaborate in the cold war and accept constraints.

Under these circumstances the role played by R. Marshak in creating the so-called Rochester Conference merits special attention and appreciation. I first met him in Rochester on my way to Princeton. He was the chairman of the department, and had also started to recruit able students from Japan. It is well known that this program has produced several illustrious physicists. I attended the second Rochester Conference, with about 20 invitees. He made a special effort to make it easy for foreign physicists to attend it, particularly the Russians. The original small local meeting soon grew and became international. Going round from U.S. to Europe and Russia, its size increased to over a thousand. The Conference was now a symbol of particle physics. Several historical discoveries were announced here. I also saw before my eyes fierce altercations between people of strong ego.

4. My discovery of spontaneous symmetry breaking (SSB) was an accidental event. But I may also say that I owe this to my education at Tokyo which had induced in me an interest in condensed matter physics, as well as to the proximity of Chicago to the birth place of the BCS theory. In my days at Osaka City University, I was attracted by the plasma theory of Bohm and Pines, and tried to generalize the many-body problem as a quantum field theory. The paper I wrote with Kinoshita at IAS was its outcome. My boss, G. Wentzel, was also interested in it, and I already knew the people at the University of Illinois: J. Bardeen, D. Pines, F. Seitz, etc. One day in 1957, R. Schrieffer, a student of Bardeen's, gave a seminar at the invitation of Wentzel, and told us about the BCS theory. The theory of Cooper pairs by L. Cooper was already known, but the BCS theory was not yet published.

My impression of the seminar was a mixture of admiration and doubt. I admired the fact that their assumed Hamiltonian led to a condensate, and formed an energy gap. At the same time, I also realized that their wave function was not an eigenstate of charge, and violated gauge invariance. How can one discuss electromagnetic properties like the Meissner effect with this theory? Shouldn't one first check gauge invariance in calculating a process involving electromagnetism? The doubt grew deeper and deeper as the success of the BCS theory became more and more conspicuous.

It took me about two years before I became convinced of the correctness of the BCS theory. In essence, I realized that a collective mode satisfied the Ward–Takahashi identity, thereby restored charge-current conservation and gauge invariance. I also found that the SSB made the electromagnetic field acquire a mass and turn into a plasmon. It is a result of mixing the two zero-modes, the Coulomb field and the Nambu–Goldstone (NG) mode. Some condensed matter physicists gave a similar account, but I was quite satisfied that I was able to clarify the issue mathematically in terms of quantum field theory.

Another point that had attracted my attention in the BCS theory from the beginning was the similarity between the Bogoliubov–Valatin quasi-fermion and the Dirac electron. The former is a superposition of states of different charge, whereas the latter is a superposition of different chirality, and the mixing leads to the energy gap formation. As soon as I resolved the gauge invariance problem, I noticed that the Goldberger–Treiman relation between the pion decay constant and nucleon mass was nothing but the Ward–Takahashi identity for chiral invariance, except that I had to make a bold assumption to ignore the pion mass. It was nice that I could now understand at least the origin of the nucleon mass.

At the 1959 Rochester Conference in Kiev there was a talk by B. Toushek about the chirality of neutrinos. After the talk I made a brief comment about my ideas, but due to the miserable condition of the Soviet Union it took two years before the conference proceedings came out. In the meantime I continued my work with a research associate G. Jona-Lasinio. After a preliminary report at a conference at Purdue University, I presented my main paper (NJL) at the 1960 Rochester Conference in Geneva. There V. G. Vaks and A. I. Larkin also presented a brief paper, but did not mention the relation to pion physics.

As I have mentioned above, there are two approaches in the methodology of physics. But my work does not belong to either of them. Instead of establishing a symmetry, it gets broken. It also breaks the axiom of the uniqueness of the vacuum. I was most afraid of objections from the second group of people who values mathematical rigor above all. Being influenced by the Sakata school, however, I am always tempted to think of concrete objects behind mathematical expressions. Thus I was fond of Dirac's interpretation of the vacuum as a sea of negative energy states. In fact, my theory was more popular among condensed matter theorists than particle theorists, I felt. As regards the N-G boson, my intension was to write a paper after collecting examples in condensed matter physics. On this point I was conscious of Tomonaga's advice not to put too many things in one paper. Heisenberg's unified theory is a bad example of this. He may have chosen his nonlinear theory as an example to show that symmetry is the guiding principle of physics, but he then claims that it is "the theory of everything". But to do that he then has to introduce contraptions like a vacuum with indefinite metric and isospin, etc. If one of them turns out wrong, the whole theory will collapse even if the others remain right or useful. Put everything in one bag, and you can lose everything. I did not want to commit the same mistake. I wanted to show the possibility of SSB in the clearest way possible under the clearest assumptions so that it can be understood by everybody. First I chose a renormalizable example like QED, but I had trouble handling higher order effects. So I changed my mind and adopted a simpler Fermi–Heisenberg type model with simple cutoff. As an added remark, Heisenberg praised my work at the 1960 Rochester Conference. It was my fond memory that thereafter I visited him in Munich, had dinner at his home, and also spoke with his collaborator K. Yamazaki.

For a few years since then I was busy applying my theory to various hadronic problems, starting the so-called soft pion physics working with research associates Y. Hara and others. My paper on the loop expansion of the S-matrix is also related to it. In the meantime the SU(3) symmetry was discovered. The quark model made its appearance. The NJL model treats the nucleon as the fundamental field, but it can also be used as an effective theory of QCD by substituting nucleons with quarks. In the 1960s I was occupied with strong interactions as the first problem to resolve, and did not think of weak interactions yet. It was not clear either how to apply the theory of gauge fields to strong interactions.

I will add a story of what happened since then. The BCS theory can be rewritten as an effective theory of the Landau–Ginzburg–Higgs (LGH) type, so it also contains, besides the NG boson, a "Higgs" boson having twice the fermion mass. It is the sigma meson in the case of hadrons. But in the case of superconductivity, I did not realize this until told by C. M. Varma. I then studied this and discovered that the same relation holds in the collective mode of Helium superfluid. The approximate 1:2 ratio is a kind of supersymmetry which I called quasi-supersymmetry. As in the case of supersymmetry, one can also factorize the Hamiltonian. So I decided to investigate all BCS type phenomena systematically. In fact the BCS theory was already used to interpret the pairing energy. I directed my attention to the Interacting Boson Model (IBM) of

Arima and Iachello. It is more complicated than superconductivity or superfluidity, but I found that one of the symmetry breaking chains can be interpreted as SSB. However, my proposal did not attract much attention by the nuclear physics community because I was not able to make precise predictions.

The next problem I took up was the more important one, namely the interpretation of the Higgs field. In the Standard Model it plays the role of giving mass to the W and Z bosons as well as to the fundamental fermions. But it cannot make predictions about the latter. They are mere adjustable parameters. I wanted to imagine some dynamics behind the Higgs field like the BCS theory. Naturally it gave rise to ideas like top condensation. The Higgs boson being a scalar, its exchange between fundamental fermions cause an attraction, so it plays the role of the phonon in superconductivity, causing SSB which in turn leads to the formation of a bound state, the Higgs boson. It is a kind of bootstrap mechanism Higgs $\rightarrow$ SSB $\rightarrow$ Higgs mediated by the top quark. This was my idea. W. Bardeen wrote it in the form of NJL and made more precise calculations. The Higgs mass turned out somewhat lower but did not agree with experimental indications. Besides, the problem of the masses of other fermions still remained.

5. Again I will return to the 1960s. Gauge invariance was my concern in the case of the BCS theory. On the other hand, it was the problem of statistics in the case of the quark model, where the quark is treated as a boson in hadronic phenomenology. On this point Gell-Mann was not very clear, saying that the quark may be a mere mathematical symbol. I was not satisfied with this. O. Greenberg proposed to solve the problem in terms of statistics of order three, but it was again too formal. So I thought of increasing quark's internal degrees of freedom. As a minimum I started with two, but the result was not very transparent. When I ventured to three, there emerged a beautiful symmetry; all hadronic states could be interpreted as SU(3) singlets. T. Takao, Y. Miyamoto and A. Tavkhelidze also published similar ideas but they were different in details.

As soon as my reprint was sent out, a modified one with some group theoretical improvements reached me from M. Y. Han, a post-doctoral research associate at Syracuse University. In this way our joint paper was born. There I postulated a superstrong force, to be represented by an SU(3) (or three-color) gauge field. But I was completely in the dark about how this could be realized. In those days the non-Abelian gauge theory had not yet been firmly established. Subsequently I noticed that the SU(3) degree of freedom could be used to assign integer charges to all quarks. But this assignment is not SU(3) invariant. To remedy this I used a QED analogy to show that the SU(3) singlet had the lowest energy. At any rate, the color degree was introduced for theoretical reasons. Then phenomenologically hadrons could also have color. As the number of new hadrons grew, however, there was no indication for the flavor SU(3) classification to break down. It was natural that my idea was ignored by the phenomenologists.

These efforts prompted me to study group representations in general. It was partly for reasons of general interest, but also because I wanted to find some order in hadronic spectroscopy. The linearly rising Regge trajectory seemed to suggest some dynamical symmetry like the SU(4) symmetry in hydrogen atom. In fact wave equations involving multiple spins or masses had been studied by many people. Dirac, for example, used an infinite dimensional representation of the Lorentz group. As a first exercise I found that its typical unitary representations can be constructed in terms of the familiar creation and annihilation operators, and found a spectrum of the inverse hydrogen type. Later I learned that this was a generalization of the equation discovered by E. Majorana in 1932. I also examined the nonrelativistic hydrogen atom, and was able to express the Schrödinger equation in terms of creation and annihilation operators. As it happens, similar programs were being pursued by A. O. Barut, C. Fronsdal, and T. Takabayashi independently, so I kept close touch with them for two years or so.

The program of unitary representations, however, soon collapsed. As it turned out, unitary representations of the Lorentz group are in general afflicted with various diseases such as violation of CPT, spin-statistics, causality, existence of tachyons; unnatural mass spectra and g-factors. Some of the troubles can be avoided by a judicious choice of representation, but a general principle is lacking. After all, the nature seems to have adopted the local field theory. It is not possible to pick only a part of the properties of nature and try to describe it in a simple way, it seems. This was also what I experienced in my days at Osaka City University in the case of the "Bethe–Salpeter equation" written in a differential form (in ladder approximation). Although it looked simple and elegant, I did not pursue it because of the existence of unphysical solution. So I decided to give up this time, too.

It is ironic that the next year, i.e. 1968, the Veneziano model made its appearance. It is not field theory, but a realization of the duality principle derived from Chew's idea of bootstrap Regge trajectory. I was immediately attracted by it although I had just given up on the mass spectrum problem. I wondered what physics lay behind his formula. The first thing to do was to decompose it into a sum of Breit–Wigner resonances, determine their multiplicity, and to check if the residues are positive. I started to work on this with P. Frampton, and soon found that the residues were positive at least asymptotically. The next task was to find a general algorithm for this decomposition. The Veneziano formula represents two-body scattering expressed by a beta function. While I was playing with its integral representation of Koba and Nielsen, one day I expanded the second factor into a power series:

$$B(-s+a,-t+b)=\int_0^1 x^{-s+a-1}(1-x)^{-t+b-1}dx,$$

$$(1-x)^{-t+b-1}=\exp\left[(-2p\cdot p'+2m^2-b+1)\Sigma\frac{x^n}{n}\right].$$

If $2m^2-b+1=0$, each term is a function of p and p'. The factor $1/n$ is reminiscent of the energy denominator $1/E$ that appears in the Fourier expansion of Green's function. This means $E\approx n$, namely a one-dimensional harmonic oscillator, isn't it? Thus the idea that the hadron is like a string was born. In order to account for the constant terms, however, it was necessary to introduce extra dimensions like in the Kaluza–Klein theory. I thought this to be unnatural, and did not make further progress for a year or so.

In the summer of 1970, a Rochester Conference was held in Kiev. I also got an invitation to speak at a summer school in Copenhagen. I decided to leave my family with a friend in California before flying to Europe. But our car broke down while driving through the Great Salt Lake Desert in Utah. We had to call help from Salt Lake City to fix the engine, and stay for three days in a town called Wendover in the desert. (Years later J. Cronin set up an experimental project called CASA in a nearby desert to detect cosmic gamma rays.) I missed the flight to Europe, and had to spend the summer in California with my family. I had sent my lecture note to Europe already. Being only a note, I casually wrote down a worldsheet action (the Nambu–Goto action), and trusted that it will be published in the proceedings. But the proceedings never came out. I am grateful to the good will of people who kindly quoted the unpublished note.

6. The 1960s were the time of clarifying the properties of strong forces. We had reached Taketani's second stage, and a glimpse had appeared on the final stage of the Standard Model. The properties of quark dynamics could be gleaned from deep inelastic scattering, Feynman's parton model and the like. But there was no further decisive step. Toward the end of the decade, people were generally in a pessimistic mood. I was driven to this mood not only because of the unpopularity of the three-color theory and the failure of infinite component equations, but also I was greatly influenced by the social and political turmoil of the period. The latter affected my personal life, and I began to fear for my security. As the Vietnam war escalated, our research condition tightened. Budget cuts caused massive unemployment of scientists. Rumor was circulating of PhD cab drivers. R. Marshak, then president of American Physical Society, asked A. Wightman and me to do an unemployment survey among fresh PhDs. The result was a gloomy one, which affected me, too. That was the year of the accident in Utah.

Thereafter I made some progress. It was clear that strings can confine quarks. On the other hand, I also had a theory of observable integer charge quarks. Which of them should I favor? In the end, however, the string theory had hit a wall as the fundamental theory of hadrons. So reported K. Kikkawa to me as a conclusion reached at an Aspen meeting. That was because string theory was not a local field theory, as in the case of infinite component equations. Around just that time there occurred the discovery of asymptotic freedom which replaced string theory to resolve the problem of quark confinement by QCD, an orthodox quantum field theory. Thus string was an approximate description of one aspect of strong forces.

Particle physics began in the 1930s and took 40 years to reach maturity in the 1970s. QCD as a theory became firmly established, and together with the Salam–Weinberg theory formed the so-called Standard Model. The appearance of J/ψ and D meson were discovered and confirmed the existence of the second generation of quarks and leptons expected by the works of Cabibbo and the Sakata group, and immediately this was followed by the third generation as predicted by Kobayashi and Maskawa. The fundamental laws governing particle physics is essentially complete. From now on our task will be to check the Standard Model in more detail. If disagreements are found, it will then signal the beginning of the next cycle, as is expected or hoped for by everybody. These tests were already in progress.

More noteworthy, however, is the dramatic change of the character of particle physics that has taken place. It may be said that the 1970s were a period of phase transition. Throughout the history of physics, theory and experiment walked hand in hand. But now the Standard Model was immediately followed by supersymmetry and supergravity, and theory started to walk alone. Its character has become mathematical and formal. People trained at school in phenomenology cannot follow it and drop out. (A similar phenomenon must have occurred after relativity and quantum mechanics.) At the same time, theory and experiment joined hands to start a rapid growth together in cosmology and astronomy. Physics is going to cover the entire range from the ultimate small to the ultimate large in a unified way. Another change is that particle physics has become big science. Research is now a collective enterprise both in theory and experiment. A good example is superstring, as you can see in the great number of papers that appear on the internet everyday. All these circumstances also seem to have led to a slower pace of progress than in preceding years.

Reference

[1] Translated from Bulletin of *Jap. Phys. Soc.* **57**, No. 1, 2003. For the history of particle physics, see, for example, *Prog. Theor. Phys. Suppl.* No. 105 (1991); V. L. Fitch and J. Rosner, *Twenty Century Physics* (Amer. Inst. Physics, 1995), p. 665.

Editor's Note

This paper by Nambu on his reminiscences, in English, is the very last work of his life. The paper was originally published in 2003 in Japanese only (*Bulletin of Japanese Physical Society*, Vol. 57, No. 1).

As Professor Tohru Eguchi and I (Moo-Young Han) were preparing the publication of the book *Nambu: A Foreteller of Modern Physics* (World Scientific, 2014), we wanted very much to have the Japanese article translated into English and included in the book. When we mentioned this to Nambu, much to our surprise and delight, he volunteered to do the translation himself! That was in the summer of 2013. Despite the fact that Nambu, then at the age of 92, was recovering from the first stroke he suffered earlier and was receiving dialysis three times a week, he finished the translation in no time.

Soon after our book was published in the spring of 2014, Nambu suffered a second stroke from which he never recovered and finally passed away on July 5, 2015 at the age of 94. And that is how this paper turned out to be the very last work of Nambu.

Printed in the United States
By Bookmasters